MEDIA AND COMMUNICATIONS - TECHNOLOGIES, POLICIES AND CHALLENGES

VIDEOCONFERENCING: TECHNOLOGY, IMPACT AND APPLICATIONS

MEDIA AND COMMUNICATIONS – TECHNOLOGIES, POLICIES AND CHALLENGES

Videoconferencing: Technology, Impact and Applications
Adam C. Rayler (Editor)
2010. ISBN: 978-1-61668-285-9
2010. ISBN: 978-1-61668-407-5 (E-Book)

Tapping the Television White Space Spectrum: A Revolution in Public Airwave Use
Julie E. Hendriks (Editor)
2010. ISBN: 978-1-60692-956-8
2010. ISBN: 978-1-61728-439-7 (E-Book)

Digital Books: Competition and Commerce
Oliver A. Hagen (Editor)
2010. ISBN: 978-1-61728-299-7
2010. ISBN: 978-1-61728-450-2 (E-Book)

Information and Communication Technologies Policies and Practices
Almas Heshmati and Sun Peng (Editors)
2010. ISBN: 978-1-60876-671-0

Intelligibility Research and Communication Issues in Emergency Situations
Samuel A. Fletcher (Editor)
2010. ISBN: 978-1-61668-634-5
2010. ISBN: 978-1-61668-732-8 (E-Book)

Video Encryption Technology and Application
Zhengquan Xu and Jing Sun
2010. ISBN: 978-1-61668-331-3
2010. ISBN: 978-1-61668-734-2 (E-Book)

Reality Television – Merging the Global and the Local
Amir Hetsroni (Editor)
2010. ISBN: 978-1-61668-315-3

Noise-Induced Hearing Loss in Youth Caused by Leisure Noise
Hannah Keppler, B. Vinck and I. Dhooge
2010. ISBN: 978-1-61668-200-2
2010. ISBN: 978-1-61668-460-0 (E-Book)

Mediating Health: The Powerful Role of the Media
Deborah Begoray, Mimi Cimon and Joan Wharf Higgins
2010. ISBN: 978-1-61668-324-5
2010. ISBN: 978-1-61668-614-7 (E-Book)

Spectrum Issues for the New Communications Age
Caroline D. Underwood (Editor)
2010. ISBN: 978-1-61668-544-7
2010. ISBN: 978-1-61668-725-0 (E-Book)

The Newspaper Industry and Journalism in Transition
Rebecca E. Greene (Editor)
2010. ISBN: 978-1-61728-166-2
2010. ISBN: 978-1-61728-323-9 (E-Book)

MEDIA AND COMMUNICATIONS - TECHNOLOGIES, POLICIES AND CHALLENGES

VIDEOCONFERENCING: TECHNOLOGY, IMPACT AND APPLICATIONS

ADAM C. RAYLER
EDITOR

Nova Science Publishers, Inc.

New York

For permission to use material from this book please contact us:
Telephone 631-231-7269; Fax 631-231-8175
Web Site: http://www.novapublishers.com

LIBRARY OF CONGRESS CATALOGING-IN-PUBLICATION DATA

Available Upon Request

ISBN: 978-1-61668-285-9

Published by Nova Science Publishers, Inc. † New York

CONTENTS

PREFACE

A videoconference is a set of interactive telecommunication technologies which allow two or more locations to interact via two-way video and audio transmissions simultaneously. It has also been called 'visual collaboration' and is a type of groupware. It differs from a videophone call in that it is designed to serve a conference rather than individuals. This book explores the technological advances that have extended the use of videoconferencing systems in a variety of areas; such as education, health and medicine, business, law, security and private use.

Chapter 1 - In the last decade, there has been an explosive growth in mobile computing technology and obviously an increase demand for platforms with multimedia application support. High speed Internet connectivity has become more widely available at a reasonable cost and the cost of video capture and display technology has decreased. Consequently personal video-conference systems based on a webcam, personal computer system, software compression and broadband Internet connectivity have become affordable for the general public. Also, the hardware used for this technology has continued to improve in quality, and prices have dropped dramatically. The availability of freeware (often as part of chat programs) has made software based video-conferencing accessible to many. The desire to be connected "*any-time, any-where, and any-way*" has led to an increasing array of heterogeneous systems, multimedia applications, devices, and service providers. There has been an evolution of the communication systems topologies towards *user-centric* frameworks, in which dynamic quality of service (QoS) reconfiguration constitutes a critical issue yet to be addressed.

Here, the impact (i.e. motivation and operating guidelines) of QoS assessment in video-conferencing applications and real-time interactive multimedia communications is discussed and investigated. We consider the end-to-end service quality methodologies presenting a QoS evaluation technique by means of an unconventional use of tracing watermarking. The rationale behind the approach is that the alterations suffered by the watermark are likely to be suffered by the data, since they follow the same communication link. We first focus on embedding techniques in transformed domains (e.g. Fourier and Wavelet) and then, we concentrate on 3G video-conferencing communication scenarios, where each customer transmits both voice and MPEG-4 video sequences in real time. Typical video-phone call consists of close shots of the speaker, framing his/her face in the foreground. We propose and adopt a novel colour space (*YST*) to minimize the perceptual distortions on face coding introduced by the tracing watermarking. The results show the benefits obtained in digital watermarking by the new representation versus the conventional approach. In fact, the sensitivity of the *YST* representation outperforms the conventional one in terms of objective (mean square error, peak signal-to-noise ratio) and subjective (video quality metrics) indicators.

Our results confirm the validity of such approach. The advantages of distributing the management process are (i) an easier and more precise localization of QoS problems, (ii) a better knowledge of local situations, (iii) a lower complexity for QoS agents and (iv) an increase in possible actions.

Chapter 2- Although the technology to support site-to-site videoconferencing has been practicable for over 70 years (Fullwood, 2007), video-mediated communication (VMC) has failed to shape the social world to the same extent as the telephone. Up until recently, the corporate sector has been the dominant consumer of VMC. It is only in recent years that video-mediated communication has penetrated the personal sector on a global scale. For over a decade now, Internet Protocol based videoconferencing has provided the general public with a reasonably priced alternative to more conventional means of distanced communications. VMC is an emerging market and is likely to continue to grow as technology advances. In a survey of 1,450 respondents, Rainie (2005) found that 16% of American adult Internet users had used a web cam to view another person or place. This translates into approximately 21 million Americans who had utilised the Internet for VMC purposes in 2005. Currently, there are estimated to be around 251 million Americans accessing the Internet, approximately 74.4% of the total population (Internet World stats, 2009), so we would expect this figure to have increased significantly. Rainie suggests that the 2005 figures represent an increase in web cam usage from previous years and this can be accredited to two specific factors: web cams are now cheaper and easier to set up, and broadband speed makes it easier to display images. As video-mediated technologies become a more pervasive aspect of our daily lives, questions relating to the psychological impact of using the technology become ever more prevalent.

There are many advantages that VMC should offer over the telephone. For example, VMC presents the opportunity of duplicating many of the benefits associated with interacting in a co-present environment (Sellen, 1997). According to Campbell (1998), videoconferencing systems closely imitate face-to-face meetings, without encumbering the user with the financial and physical costs of travelling. Unlike audio-only communications, VMC permits participants to perceive the reactions of their colleagues. According to the 'non-verbal hypothesis' (Whittaker, 1995; 1996), access to non-verbal cues should allow users to respond with greater comprehension and sensitivity. For example, one should be able to notice if another person is smiling or frowning (Campbell, 1998), and this may have a profound impact on the exchange. Furthermore, Isaacs and Tang (1993) indicate that adding a visual channel to remote communication produces many benefits for users. These include: allowing listeners to signal their understanding (or lack of) in the non-verbal channel; speakers can supplement verbal information with gestures; listeners can respond to speakers in an entirely visual capacity (for example, head-nodding); attitudes can be communicated with posture and facial expressions; and the significance of conversational pauses can be interpreted more easily. For all of these reasons, one would expect video-mediated communication to be a more satisfactory, effective and efficient mode of interaction than audio-only communication.

Chapter 3- Technological advances have extended the use of videoconferencing systems in a variety of areas, such as education, health and medicine, business, law, security, private use, etc. The health and medicine area in particular strives at improving clinical attention to patients by implanting videoconferencing into specific devices (equipped rooms) and general devices (systems installed in personal devices) so as to improve the clinical attention that is given to patients. Videoconferencing is implanted in telemedicine, teleassistance, telesurgery, etc. in order to help sanitary staff with the elaboration of diagnoses, assistance to patients in rural areas, the interactive exchange of clinical information between specialists, assistance in long-distance operations, etc.

Videoconferencing requires the presence of an equipped room or a device (PC, PDA, mobile phone, etc.) that contains the necessary hardware to support it (video camera, microphone and speaker, specific software). The videoconferencing systems that are used in the medical/sanitary field are similar to those of other environments, except for the presence of peripheral devices such as medical images devices, microscopes, monitoring devices, etc, which may require additional signal transmission.

In many cases, maintaining a certain quality of image and sound in real time is essential, and therefore very high bandwidth becomes indispensable. For this reason, videoconferencing systems use compression formats that can be adapted to the existing transmission means while maintaining an acceptable quality of image and sound (service quality). Also, these systems must follow audio/video standards that make it possible to implant them into various systems and adapt them to specific tasks.

The videoconferencing standards that are used in the fields of health and medicine are mainly the following: H.320, a standard that is designed for ISDN communications and groups the H.261 norm for video, G.711/G.722/G.728 for audio, and T.120 for data; and H.323, a standard that is designed for IP communications and groups the H.261/H.323 norms for video, G.711/G.722/G.723.1/G.728 for audio, and T.120 for data.

Chapter 4- Information and Communication Technologies (ICT) and especially advanced learning technologies of synchronous transmission can effectively support the educational uses of Video Conferencing features by providing innovative learning and instructional strategies that would improve communication, collaboration and interaction between the learner and the educator. The last 10 years have developed in Greece and Cyprus several important instructional applications via videoconferencing. The objective of this chapter is to present the main educational videoconferencing K-12 applications in Greece.

Chapter 5- The development of quality and user satisfying videoconference applications at a low cost has been hindered in the past by, among other reasons, low quality connections and a difficulty for the average user to establish end-to-end connections without hassle. In this chapter, we describe some of the latest methods and developments that overcome these problems. In particular, we discuss ways to make establishment of universal videoconference connections easier by overcoming the problem of participants that reside behind NAT routers, by deploying new protocols at the application or even the network layer. An example of the former case is the development of establishment protocols specifically designed for this purpose, while an example for the latter case is the deployment of IPv6, which aims to make NAT deployments obsolete. Furthermore, we discuss efforts to provide targeted support for quality of experience in networks that support some kind of traffic classification through the usage of dynamic mechanisms and dynamic network re-configurations. Apart from describing the state of the art in the aforementioned areas, we provide detailed insight in specific research efforts that have taken place and assess their results in the overall user experience.

Chapter 6- Web-based videoconferencing technologies hold far-reaching potential to increase and improve delivery of evidence-based substance abuse treatment interventions. A review of the literature reveals very few reports on the efficacy of web-based videoconferencing intervention strategies for treating substance use disorder, and many of those have been descriptive reports that do not meet the requirements of establishing effective and replicable evidence-based care. Comparatively more studies have reported on the use of computer-administered pre-programmed treatment and education modules. These approaches can be very helpful, but do not expand the availability of flexible and professional counseling services for patients in long-term therapy relationships that can be realized with web-based videoconferencing. One of the next important advances in the substance abuse treatment continuum is to evaluate how web-based videoconferencing can help engage patients with serious and chronic problems in the long-term outpatient treatment process. Patients requiring long-term treatment often require responsive and flexible levels of counseling to manage symptoms that fluctuate greatly in scope and intensity over time. Unfortunately, the overall effectiveness of long-term therapy is often seriously hindered by treatment demands to on-site scheduled counseling and psychosocial services that often continue for many years. These demands reduce utilization of therapy services with resulting negative impacts on treatment outcome. Web-based videoconferencing service delivery provides a remarkable opportunity to reduce some of the inconvenience of counseling and so facilitate the long-term treatment process while at the same time allowing pursuit of other responsibilities (e.g., child care) or treatment goals (e.g., employment) essential to rehabilitation. A recent study by our research group demonstrated both increased satisfaction and improved attendance in patients receiving long-term substance abuse treatment who were assigned to web-based counseling versus on-site counseling, with no differences in treatment outcomes. However, much more work is needed to establish how best to use this technology since the great majority of published studies have focused on individuals with mild to moderate problems and/ or on brief interventions. Using web-based videoconferencing to expand the continuum of care holds great promise to improve treatment satisfaction, adherence and retention for the many patients with chronic substance use problems requiring long-term outpatient substance abuse treatment.

Chapter 7- The advanced learning internet technologies indicated a need for methodologies, tools and pedagogies that support the introduction of synchronous e- learning applications and especially Interactive Videoconferencing (IVC) into the whole learning environment. The objective of this chapter is to present the theoretical background and the methodology that have been developed, to implement a pedagogical model for IVC instruction. The proposed pedagogical model constitutes the results experienced from an IVC project in the University of Crete during the last seven years, under the title `ODUSSEAS', which took place between eight hundred students and thirty teachers in eighteen elementary schools in Greece and Cyprus.

Chapter 8- The organization of a successful VC requires detailed design and preparation, as there are a number of factors to take into consideration. Depending on the VC type, the participants' needs and the set goals we should investigate aspects like the necessary technological equipment, the lay-out of venues (Anastasiades, 2007; Mouzakis et al., 2004), the organizational arrangements and, last but not least, the conditions for a warm and collaborative environment for the participants (Anastasiades, 2008).

In: Videoconferencing: Technology, Impact and Applications ISBN: 978-1-61668-285-9
Editor: Adam C. Rayler, pp. 1-33

Chapter 1

IMPACT OF QUALITY OF SERVICE ASSESSMENT IN VIDEO-CONFERENCING: TECHNOLOGIES AND APPLICATIONS

F. Benedetto*[*] *and G. Giunta

Digital Signal Processing, Multimedia, and Optical Communications Lab.,
Dept. of Applied Electronics, University of ROMA TRE,
Via della Vasca Navale 84, 00146 Rome, Italy

ABSTRACT

In the last decade, there has been an explosive growth in mobile computing technology and obviously an increase demand for platforms with multimedia application support. High speed Internet connectivity has become more widely available at a reasonable cost and the cost of video capture and display technology has decreased. Consequently personal video-conference systems based on a webcam, personal computer system, software compression and broadband Internet connectivity have become affordable for the general public. Also, the hardware used for this technology has continued to improve in quality, and prices have dropped dramatically. The availability of freeware (often as part of chat programs) has made software based video-conferencing accessible to many. The desire to be connected "*any-time, any-where, and any-way*" has led to an increasing array of heterogeneous systems, multimedia applications, devices, and service providers. There has been an evolution of the communication systems topologies towards *user-centric* frameworks, in which dynamic quality of service (QoS) reconfiguration constitutes a critical issue yet to be addressed.

Here, the impact (i.e. motivation and operating guidelines) of QoS assessment in video-conferencing applications and real-time interactive multimedia communications is discussed and investigated. We consider the end-to-end service quality methodologies presenting a QoS evaluation technique by means of an unconventional use of tracing watermarking. The rationale behind the approach is that the alterations suffered by the watermark are likely to be suffered by the data, since they follow the same

[*] Corresponding author: E-mail: fbenedet@uniroma3.it

communication link. We first focus on embedding techniques in transformed domains (e.g. Fourier and Wavelet) and then, we concentrate on 3G video-conferencing communication scenarios, where each customer transmits both voice and MPEG-4 video sequences in real time. Typical video-phone call consists of close shots of the speaker, framing his/her face in the foreground. We propose and adopt a novel colour space (*YST*) to minimize the perceptual distortions on face coding introduced by the tracing watermarking. The results show the benefits obtained in digital watermarking by the new representation versus the conventional approach. In fact, the sensitivity of the *YST* representation outperforms the conventional one in terms of objective (mean square error, peak signal-to-noise ratio) and subjective (video quality metrics) indicators.

Our results confirm the validity of such approach. The advantages of distributing the management process are (i) an easier and more precise localization of QoS problems, (ii) a better knowledge of local situations, (iii) a lower complexity for QoS agents and (iv) an increase in possible actions.

I. Introduction

In recent years, there has been an explosive growth in wireless/mobile networks and obviously an increase demand for platforms with mobile multimedia application support. The commercial success of the second-generation (2G) mobile networks and the growth rate of the Internet over the last decade have led the Information and Communication Technology (ICT) industry and research community to focus their effort towards an integrated framework that will provide a wider range of services to mobile users [1]–[3]. Moreover, there is an increased need to provide added-value services to customers of third-generation (3G) mobile networks. Progress in Telecommunications and Networking is fostering the development of high-speed and ubiquitous networks, both wired and wireless (i.e. heterogeneous configurations), characterized by an unprecedented degree of transport capacity and flexibility. Open-service framework concepts will enable the provisioning of a wide spectrum of services merging voice and data features. It offers potential benefits to shareholders by enabling application services to make better use of network functions (and thereby retain more of the services' value-add within the network).

However, the lacks of access and location transparency as well as of re-configurability and adaptability capabilities have proven to be a major shortcoming for the evolution towards a universal 4G mobile wireless network [3]. These impediments which result in limited user friendliness in contrast to the open easy-to-access and easy-to-use Internet paradigm are motivating a research effort to alleviate the above-mentioned shortcomings prior to the commercial launch of all variations of 3G systems. Therefore, the future trend is going to provide the transmission rates required by most potential users, but the practical question arises of *what* type of services can be performed, as well as *when* and *where* [4]. There exists a clear trend for the convergence of IP protocol to wireless, that is, what we now call wireless IP, leading to the Wireless Internet [5]-[6]. Innovating mobile multimedia services is going to generate the need for new service enablers, as well as new data processing and managing techniques. This trend will drastically change the topology of the communication systems:

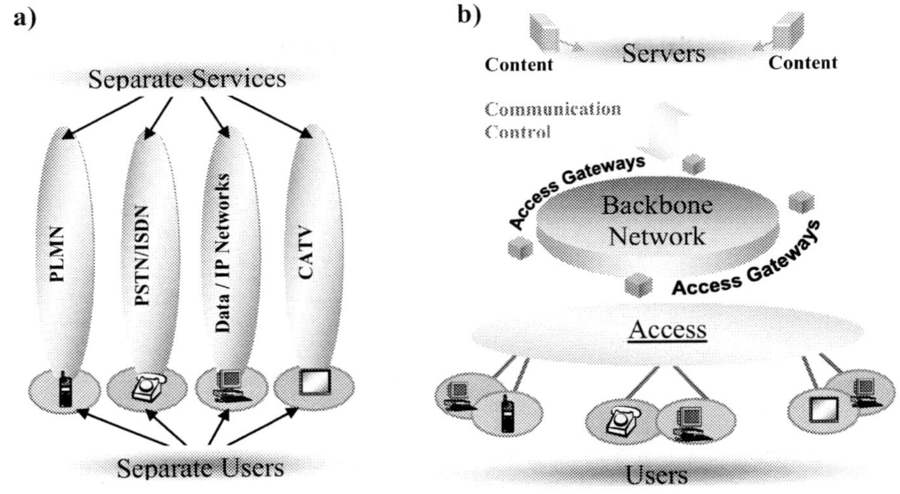

Figure 1. Mobile multimedia services will drastically change the topology of the communication systems: the network configurations of today (a) evolve towards the network of tomorrow (b).

- from the configuration of today (see Figure 1a), characterized by separate networks, separate users and separate services;
- towards the next configuration networks (see Figure 1b), characterized by separate accesses, same core network, same user on different accesses and same services.

This means an evolution from a *device-centric* approach to a *user-centric* design, i.e. an approach whereby the user preferences rule [7]. One remarkable difference between the two approaches is that user-centric management schemes tend to distribute qualities of services (QoS) more evenly to users than device-centric ones [8]. Therefore, to practically realize such a user-centric design, we need to explore the concept of QoS.

In this chapter, we focus on the end-to-end service quality methodologies applying a QoS evaluation technique by means of an unconventional use of fragile watermarking. This technique (recently introduced in [9]-[11] with initial promising results) is here widely applied to typical operating scenarios of wireless/mobile communications through extensive simulation trials. In particular, we show that the receiving mobile station is able to verify whether the negotiated quality level has been actually maintained by the effective link or not by means of this QoS measurement. Such a QoS index can be usefully employed for a number of different purposes in wireless multimedia communication networks such as: control feedback to the sending user on the effective quality of the link; detailed information to the operator for billing purposes and diagnostic information to the operator about the communication link status [4]. This chapter is organized as follows. Section II is about the impact of QoS assessment in mobile video-conferencing applications for wireless networks, showing motivations and operating target of the QoS in a *user-centric* framework. Section III discusses the tracing watermarking technique, first in the discrete cosine transform (DCT) domain then in the discrete wavelet transform (DWT) domain. Comparisons and results in wireless video-communications close the discussion of Section IV. The importance of color for the QoS assessment in video-conferencing application is highlighted in Section V, while Section VI shows the importance of choosing a proper color space for the signal processing technique under investigation. Section VII introduces a new color space, named YST, for the tracing watermarking. The Section ends with the application of this novel color space to wireless mobile devices for video-conferencing purposes. Finally, a discussion about dynamic QoS management in 3G communications for application in a shared environment concludes this chapter.

II. Impact of QoS Assessment in Video-Conferencing Application

The problem of 'matching' the application to the network is often described as a "quality of service" (QoS) problem [17]. The problem is twofold: there is the QoS required by the multimedia application, which relates to visual quality perceived by the user, and the QoS reachable by the transmission channel and the network, which depends on the physical link's capabilities [4]. The key parameters for QoS requirements about transmission are the mean bit rate and the variation of the bit rate [17]. The mean rate depends on the compression algorithm as well as the characteristics of the source video (such as frame size, number of bits per sample, frame rate, amount of motion, etc.). Many video coding techniques include a certain degree of compression tuning that bound of the mean rate after encoding. However, the achievable mean compressed bit rate is actually limited for a given source (with a particular frame size and frame rate). Real-time video is sensitive to delay by nature.

The QoS requirements in terms of delay depend on whether video is transmitted one way (e.g. broadcast video, streaming video, playback from a storage device) or two ways (e.g. video conferencing) [17]. The original temporal relationships between frames have to be preserved in simplex (one-way) video transmission because any associated media that is 'linked' to the video frames should remain synchronized, but a certain degree of congruent delay is admitted. The most common example is accompanying audio, where a loss of synchronization of more than about 0.1 *s* can be obvious to the viewer [17]. Duplex (two-way) video transmission has the above constraints (constant delay in each direction, synchronization between related media) plus the obligation that the total delay from acquisition to display must be low enough. A 'rule of thumb' for video conferencing is to keep the total delay less than 0.4 *s* [17]. If the delay is longer than this, video-conversation appears like non-natural. Interactive applications also have a requirement of low delay. In such a case, an eventual too long delay between the user's action and the corresponding effect on the video source makes the application unresponsive.

Therefore, real-time QoS management represents an important requirement during mobile multimedia applications in real-time communications. The aim of QoS management is to adapt to changing conditions during runtime and to maintain committed QoS levels or to react upon changes in these levels. This contribution proposes a novel procedure to evaluate the QoS of a communication link. In practice, the receiving mobile station is able to verify whether the negotiated quality level has been actually maintained by the effective link or not. In such a case, the operator can lower the user's fare and bustle about in order to guaranty an improved service level by modifying the operating settings of the communication link. The blind operating system at the receiver of a communication link needs to decide in order to match the negotiated quality level and the effectively provided one. If the user had requested a secure (high-level) transmission, he has to be accordingly charged for the relative fare. Otherwise, the connection can be low-priced, even becoming free of charge. Moreover, like discussed in [9], it can be observed that in real-time interactive multimedia communication systems requiring error control, such as video telephony and streaming, a number of consequences follow a poor quality decision from the mobile station: first, the bit rate (and the video resolution) transmitted by the base station is lowered in a few seconds (by a higher spreading factor in the UMTS systems); second, if the mobile station declares a null quality, the link is broken down by the operator; third, frequent declarations of poor or null quality will be a reason for the operator's admission call manager to refuse the access for further calls of the same user for a given time period, i.e. until the mobile station moves into a region with higher signal-to-noise ratio (SNR). These should avoid fraudulent declarations of bad quality from the mobile station in order to transmit free of charge. In fact, the result of possibly false declarations on the QoS is to suddenly cut the communication in a few seconds, assuming the risk of no admission for further calls as it will be detailed in the following.

In the next generation networks, mobile users will have access to devices with different capabilities at different times and to various wireless access networks and priority services [3]. These requirements will lead to an evolution of the communication systems' topologies towards user-centric frameworks, in which the main goal is to transmit contents that are a best match to the user's preference and therefore most satisfying to the user [12]. Dynamic QoS reconfiguration constitutes an important constraint that is yet to be addressed in the emerging open-service architecture (OSA) framework of 3GPP. The aim of QoS management is to adapt to changing conditions during runtime and to maintain committed QoS levels or to react upon changes in these levels. The core of this paper is to obtain a QoS index in order to realize a control feedback to the sending user on the effective quality of the link, and therefore to put into action *optimal* QoS policies in terms of the effective perceived end-to-end QoS [13]-[14].

The QoS evaluation is realized by means of an operating procedure, involving the definition of a number of QoS levels and a blind assignment of the received multimedia stream to one of the possible classes of service. In practice, the receiving mobile station is able to verify whether the negotiated quality level has been actually maintained by the effective link or not. In the simplest version, the two hypotheses consist of "good" and "not-good" quality levels. The blind operating system at the receiver of a communication link needs to decide in order to match the negotiated quality level and the effectively provided one. If the user had requested a secure (high-level) transmission, he has to be accordingly charged for the relative fare. Otherwise, the connection can be low-priced, even becoming free of charge. The economical impact of the decision error is dramatic for the operator. In fact, the false alarm has a tremendous cost for the operator since it allows a user to (well) transmit free. In practice, the probability of this event has to be limited to an upper bound chosen from an economical analysis of costs/benefits. Conversely, the erroneous decision of a good quality has a direct consequence to the customer that has to pay for a too low quality call. This decision has no direct costs for the operator. In fact, it continues to charge the multimedia call at the negotiated price. In such a case, the typical user drops the call line and dials again. Nevertheless, the service inefficiency provided to the end user risks to push the customer to do not use anymore multimedia services.

Moreover, the operator must be very careful while determining the probability of such event, since the service efficiency is one of the parameters considered by the customers while they chose an operator against the concurrent ones. Therefore, the probability of such event must be minimized by the operating system. These should avoid fraudulent declarations of bad quality from the mobile station in order to transmit free of charge. In fact, the result of possibly false declarations on the QoS is to suddenly cut the communication in a few seconds, assuming the risk of no admission for further calls. Thus, by knowing the end-to-end QoS, it is possible to adopt *optimal* strategies in terms of the QoS profiling for all the active users involved in the communication [13]-[14]. Following for the watermarking setup the same methodological approach of [9], in the next Section we show some examples of application to typical wireless network scenarios in order to verify the sensitivity of the signal processing technique of [9] in such an operating environment.

III. Framework of Tracing Watermarking for QoS Assessment

In this Section, the principles of the tracing watermarking procedure for the QoS assessment are introduced. As stated before, copyright protection of multimedia contents was the original target of watermarking, but some other applications are reported in the literature (fingerprinting, broadcast monitoring, data authentication, multimedia indexing, content-based retrieval applications, etc.). Usually, these techniques use a robust watermarking which means that the embedded watermark is supposed to be detectable (i.e. the watermark used for copyright protection should not allow the use of the watermarked content until it has not been correctly extracted). Recently, an unconventional use of fragile watermarking (the watermark should not be detectable when using the watermarked content) has been proposed to blind estimate the quality of service of a video-communication [4], and for QoS provisioning and control purposes [9]-[11]. The rationale behind the approach is as follows: the alterations suffered by the watermark are likely to be suffered by the data, since they follow the same communication link.

At the receiving side, the extracted watermark is compared with its original counterpart and the watermark degradations can then be used to evaluate the alterations endured by the data, obtaining a QoS index of the communication link (as explained in principles in Figure 2). The Quality of Service is evaluated by comparing the extracted watermark with respect to the original one. In particular, its mean-square-error (MSE) is evaluated as an index of the effective degradation of the provided QoS. Thus, by knowing the end-to-end QoS, it is possible to adopt *optimal* strategies (e.g. pricing strategies) in terms of the QoS profiling for all the active users involved in the communication. In fact, such a QoS index can be usefully employed for a number of different purposes in multimedia communications such as: control feedback to the sending user on the effective quality of the link; detailed information to the operator for billing purposes and diagnostic information to the operator about the communication link status [9]. The first experimental results obtained in [4] show the capability of this unconventional use of watermarking techniques to *trace* the alterations suffered by the data through the communication channel. This quality assessment method can be applied both to uncompressed and compressed video sequences [4]. In the first case, the watermark embedding is performed on the data before coding, whereas in the second case, watermark embedding is performed on the coded bit stream.

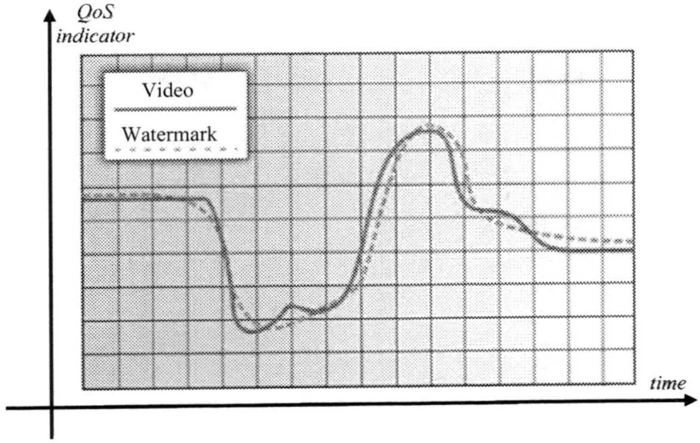

Figure 2. Idea of *tracing* watermarking: the mark degradations can be used *to trace* the alterations endured by the data.

Therefore, it can be used to estimate both the degradations introduced by the cascade coder-channel, as when the data are first coded and then transmitted through a non-ideal channel, and the ones introduced only by the channel.

a. Watermarking in the Discrete Cosine Transform Domain

More in details, the watermark (narrow band low energy signal) is spread over the image (larger bandwidth signal) so that the watermark energy contribution for each host frequency bins is negligible, which makes the watermark near imperceptible [15], [16]. Following the same methodological approach of [9], a set of uncorrelated pseudo-random noise (PN) matrices (one per each frame and known to the receiver) is multiplied by the reference watermark (one for all the transmission session and known to the receiver): $w_i^{(s)}[k_1,k_2]=w[k_1,k_2]\cdot p_i[k_1,k_2]$, where $\mathrm{w}[k_1,k_2]$ is the original watermark, $\mathrm{p}_\mathrm{i}[k_1,k_2]$ the PN matrices and $\mathrm{w}_\mathrm{i}^{(\mathrm{s})}[k_1,k_2]$ the spread version of the watermark to be embedded in the i-th frame. The embedding is performed in the DCT domain according to the following:

$$F_i^{(w)}[k_1,k_2]=\begin{cases} F[k_1,k_2]+\alpha\cdot w_i^{(s)}[k_1,k_2], & (k_1,k_2)\in S \\ \\ F[k_1,k_2], & \text{otherwise} \end{cases} \quad (1)$$

where $F_i[k_1, k_2]$ = DCT$\{f_i[k_1, k_2]\}$ is the DCT transform of the i-th frame; S is the region of middle-high frequencies of the image in the DCT domain, while α is a scaling factor that determines the watermark strength and $F_i^{(w)}[k_1, k_2]$ is the DCT of the i-th watermarked frame. In the following, the value $\alpha = 0.04$ has been always used. The i-th watermarked frame is then obtained by performing the IDCT transform of $F_i^{(w)}[k_1, k_2]$, finally the whole sequence is coded and then transmitted through a noisy channel, like shown in the block scheme of Figure 3.

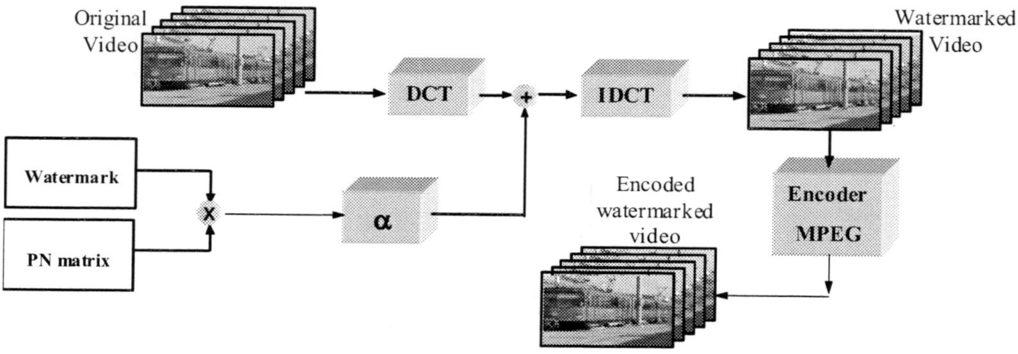

Figure 3. Principle block scheme of the watermark embedding process in the DCT domain

The receiver implements video decoding as well as watermark detection. In fact, at the same time after decoding of the video stream, a matched filter extracts the (known) watermark from the DCT of each n-th received I-frame of the sequence. The estimated watermark is matched to the reference one (despread with the known PN matrix). The matched filter is tuned to the particular embedding procedure, so that it can be matched to the randomly spread watermark only. It is assumed that the receiver knows the initial spatial application point of the mark in the DCT domain.

The Quality of Service is evaluated by comparing the extracted watermark with respect to the original one. In particular, its mean-square-error (MSE) is evaluated as an index of the effective degradation of the provided QoS. A possible index of the degradation is simply obtained by calculating the mean of the error energy:

$$E_e = \frac{1}{M}\sum_{n=1}^{M}\left(\sum_{k_1=1}^{K_1}\sum_{k_2=1}^{K_2}(w_n[k_1,k_2]-\hat{w}_n[k_1,k_2])^2\right) \tag{2}$$

where $\mathrm{w}_n[k_1,k_2]$ and $\hat{\mathrm{w}}_n[k_1,k_2]$, represent the original watermark, the extract watermark respectively, and $n = 1..M$ is the current frame index. Another video quality measure usually used for QoS assessment is the peak signal-to-noise ratio (PSNR) defined as the ratio between peak signal and root mean square (rms) noise observed between the original video and the watermarked video, where RMSE is the square root of the mean square error previously defined:

$$PSNR = 20\log_{10}\left(\frac{255}{RMSE}\right) \tag{3}$$

In particular, such QoS indexes can be usefully employed for a number of different purposes in wireless multimedia communication networks such as: control feedback to the sending user on the effective quality of the link; detailed information to the operator for billing purposes and diagnostic information to the operator about the communication link status (see Figure 4).

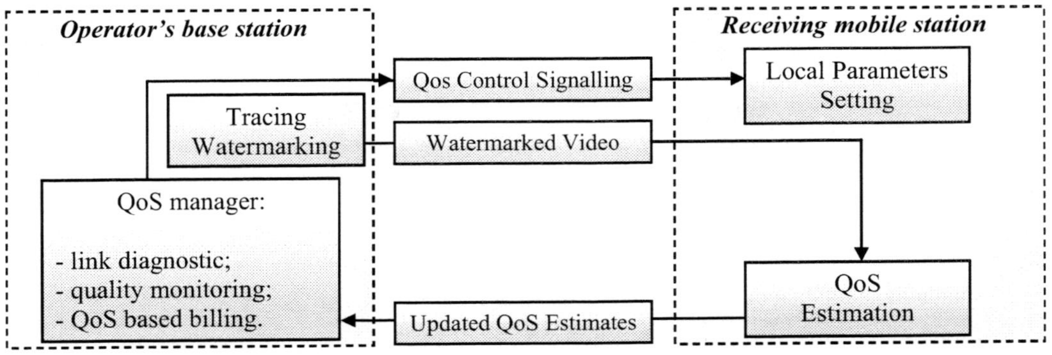

Figure 4. Principle scheme of the quality control feedback system for wireless multimedia communications

b. Watermarking in the Wavelet Transform Domain

Again (see Figure 5), a set of uncorrelated pseudo-random noise (PN) matrices (one per each frame and known to the receiver) is multiplied by the reference watermark (one for all the transmission session and known to the receiver). Like in spread-spectrum techniques, the use of different spreading PN matrices assures that the spatial localization of the mark is different frame by frame, so that the watermark visual persistency is negligible [4], [26]. Moreover, the method is robust against permanent bit errors, due to either the physical network or its management (e.g. multi-path of the transmission channel, multi-user interference, excess loading factors, etc.). After generating the marks, the embedding of the tracing mark is now performed in the DWT domain [18]-[25]. The watermark is randomized by the PN matrices and added to the DWT of each frame. After the inverse DWT, the whole sequence is coded by a video coder (either MPEG2 or MPEG4 coders) and transmitted. In practices, the watermarks are embedded in a (large) number M of the luminance I-frames.

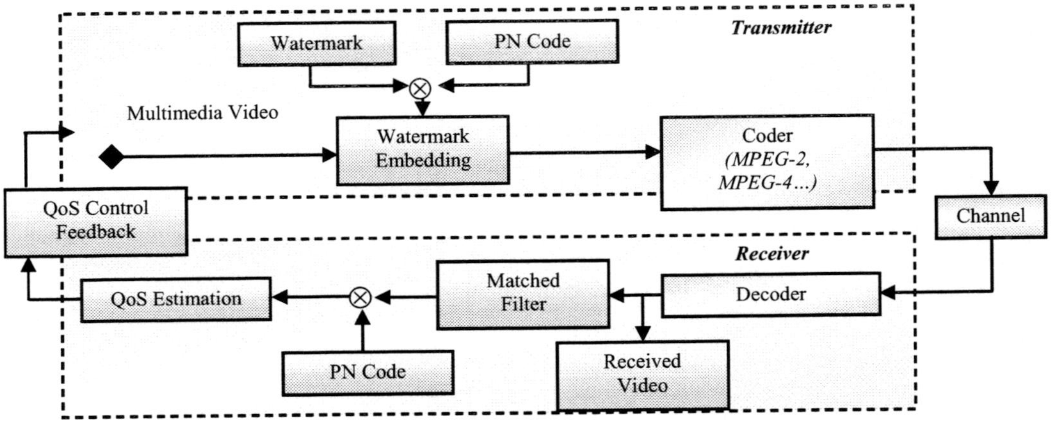

Figure 5. Block scheme of the watermark embedding and extraction processes

In this way, the watermark is affected only by the channel's errors and at the receiver side the estimation of the degradations affecting the received mark can be used to provide quality assessment of the channel. The result is that the spatial localization of the mark is different frame by frame, so that the visual latency perception is negligible and, on the other hand, it is robust against permanent bit errors, due both to the physical network or its management (e.g. multi-path of the transmission channel, multi-user interference, excess loading factors, etc.). The receiver implements video decoding as well as watermark detection. In fact, at the same time after decoding of the video stream, a matched filter extracts the (known) watermark from the DWT of each *n*-th received I-frame of the sequence. The estimated watermark is matched to the reference one (despread with the known PN matrix). The matched filter is tuned to the particular embedding procedure, so that it can be matched to the randomly spread watermark only. It is assumed that the receiver knows the initial spatial application point of the mark in the DWT domain.

The Quality of Service is evaluated by comparing the extracted watermark with respect to the original one. In particular, its mean-square-error (MSE) is evaluated as an index of the effective degradation of the provided QoS. On the other hand, this metric represent an objective indicator to evaluate the performance of the embedding process. This means that it gives poor information about the quality of the transmitted data (video sequences) really perceived by the final user (i.e. it may not be a perfect predictor of perceived quality). A metric which conversely provides subjective information about the perceived quality of the image (and of the video sequence) after the watermarking process is the video quality metric (VQM). This latter indicator reaches values near zero when degradations are imperceptible in the image; otherwise its value grows up as the image quality decreases, according to the National Telecommunications and Information Administration (NTIA) model [27]. VQM is intended to replace human evaluation with evaluation by machine. To accurately simulate human judgment, it must include some aspects of the human visual system [28]. In the next Section, we are showing the advantages obtained by the new embedding domain (DWT) as well as in terms of both objective (MSE) and subjective (VQM) indicators.

IV. Application to Wireless Video-Communications

Several simulations have been made to detect the sensitivity of digital watermark to two different degradation sources: the noisy channel affects the video quality and the co-decoder itself affects the perceived image quality. In practice, the decoding quality depends on three factors: *1)* the multiplying factor of the quantization matrix: as it increases the video quality also decreases; *2)* the bit-rate: as it increases it improves the video quality; *3)* the number of intra-coded frames.

To evaluate the sensibility of the method to the degradations due only to the channel, tests have been performed with different values of BER using the same coder quality. In all the following simulations the scaling factor regarding the watermark intensity has been set equal to 0.04, according to [9]. Increasing this value, the mark becomes more evident and a visual degradation of the video occurs. On the contrary, by diminishing its value, the mark can be easily removed by the coder and/or channel's errors. Figure 5a shows the mean square error of the decoded QCIF (176x144 pixels) "*Akyio*" video and the extracted watermark versus the bit error rate of the channel, using a MPEG-2 coder at 100 kbit/s. Figure 5b shows the MSE of the "*News*" video using a MPEG-4 coder at 100 kbit/s. Moreover, Figure 6a shows the mean square error of the decoded QCIF (176x144 pixels) "*Miss America*" video and the extracted watermark versus the bit error rate of the channel, using a MPEG-2 coder at 100 kbit/s. Figure 6b shows the MSE of the "*Foreman*" video using a MPEG-4 coder at 100 kbit/s. The above trials confirm that the watermark error is roughly proportional to the coding quality, which conversely depends on the available channel capacity. From the operating viewpoint, we are interested in the actual video over a given communication link determined by the (booked) maximum bit rate. In other words, the target quality is fixed by the negotiated channel capacity, while the actual quality depends also on the symbol errors introduced in the received data stream by the physical link, due to background noise as well as multi-path channel and interference effects. Such transmission errors are modeled as a random Poisson process, characterized by a parametric probability of symbol error (proportional to the bit error rate). Then, the actual quality needs to be detected by an in-service quality assessment measure, such as the tracing watermarking technique here discussed. For this purpose, a number of simulations have been carried out. Figure 7 reports here the PSNR of the watermarked MPEG-2 "*Miss America*" sequence (Figure 7a) and the MPEG-4 "*Foreman*" sequence (Figure 7b) versus the bit error rate of a simulated link for several (target) bit-rates. The obtained results evidence the sensitivity of the watermarking quality index to the actual quality for given target quality levels and show the capability of this unconventional use of watermarking techniques to trace the alterations suffered by the data through the communication channel. The quality assessment method presented here can be applied both to uncompressed and compressed video sequences. In the first case, the watermark embedding is performed on the data before coding, whereas in the second case, watermark embedding is performed on the coded bit stream. Therefore, the method proposed here can be used to estimate both the degradations introduced by the cascade coder-channel, as when the data are first coded and then transmitted through a non-ideal channel, and the ones introduced only by the channel. In conclusion, the presented approach allows one to blindly estimate the QoS provided by a coder/channel system without affecting the quality of the video-communications. Moreover, the proposed technique can be usefully employed for a number of different purposes in mobile multimedia communication networks such as: control feedback to the sending user on the effective quality of the link; detailed information to the operator for billing purposes and diagnostic information to the operator about the communication link status.

Let us now see the results obtained with the embedding in the DWT domain. In particular, in Figure 8, the MSE of the estimated watermark with respect to the original one versus the BER for MPEG-4 coded video sequences at different compression ratio have been considered. In particular, Figure 8 reports here the MSE of the watermark extracted from the MPEG-4 "*Carphone*" (Figure 8a) and "*News*" sequences (Figure 8b) versus the BER of a simulated link for several (target) bit-rates. It is worth noting that the MSE of the extracted watermark increases when the BER increases and the bit rate decreases. This is in accordance with the perceptual degradation that the video suffers at increasing BER and decreasing bit rate. As shown in Figure 9, the quality degradation of the watermark embedded into the host video has the same behavior of the one affecting the video, as happened before realizing the embedding in the DCT domain.

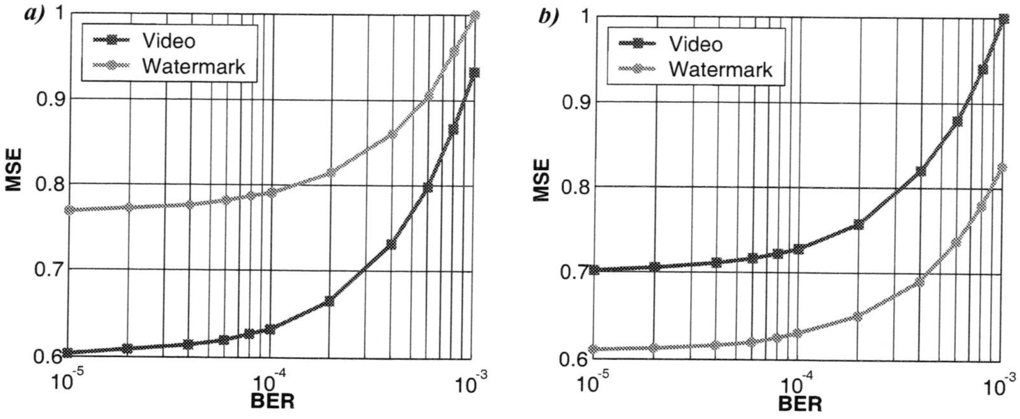

Figure 5. Watermark MSE and video sequence MSE (normalized to 1) versus the BER: for the sequence "News" MPEG-4 coded at 100 Kb/s (left) and for the sequence "Akyio" MPEG-2 coded at 100 Kb/s (right)

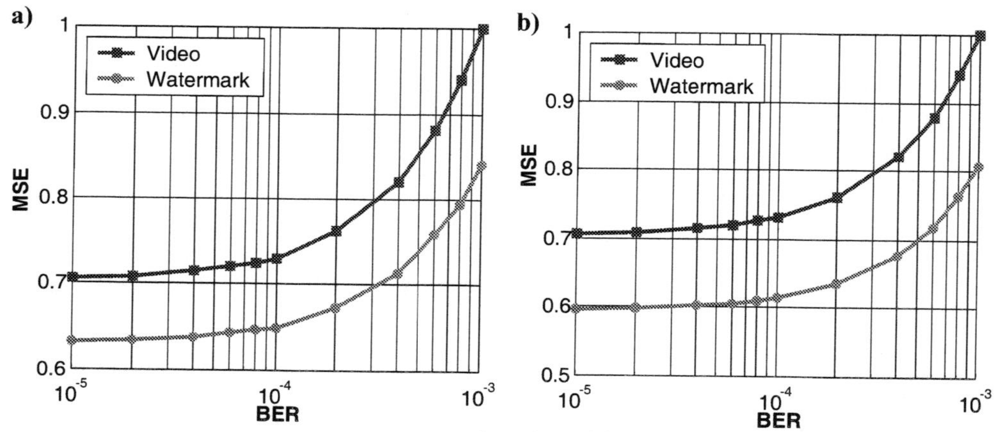

Figure 6. Watermark MSE and video sequence MSE (normalized to 1) versus the BER: for the sequence "Foreman" MPEG-4 coded at 100 Kb/s (left) and for the sequence "Miss America" MPEG-2 coded at 100 Kb/s (right)

However, the MSE represents an objective indicator to evaluate the performance of the embedding process. In order to evaluate the information about the perceived quality of the image (and of the video sequence) after the watermarking process, the video quality metric (VQM) is provided. In Figure 10, the VQM of the two MPEG-4 video sequences "*Carphone*" and "*News*" is presented. As it can be easily seen from the graphs, the curves related to the embedding in the DWT domain are always lower than the others (in the DCT domain). This means that the perceptual degradations introduced by the watermark embedding process in the image have been minimized using the DWT domain as the domain for the embedding process instead of using the DCT domain.

The obtained results evidence the sensitivity of the watermarking quality index to the actual quality for given target quality levels and show the capability of this unconventional use of watermarking techniques to still trace the alterations suffered by the data through the communication channel. Moreover, the use of the DWT domain as the watermark embedding domain introduces less artifacts and alterations in the watermarked video sequence.

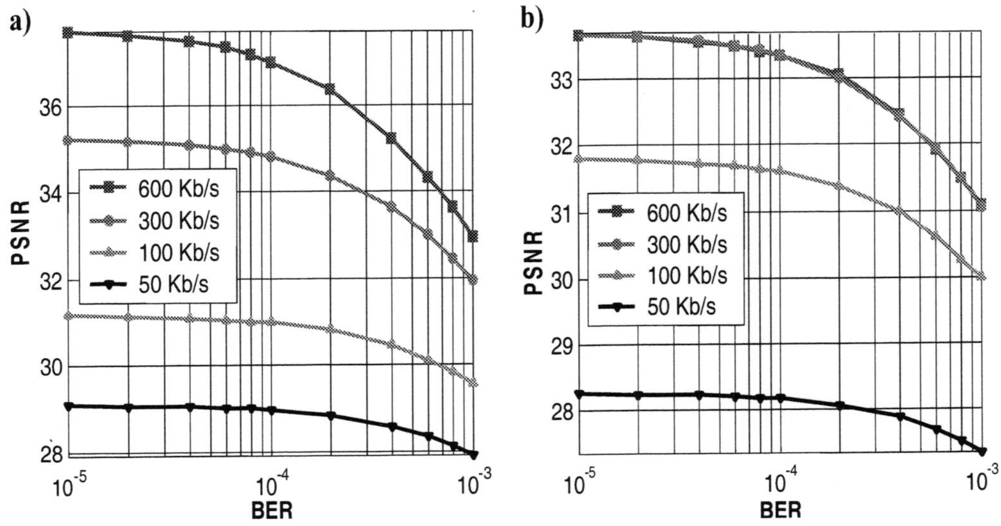

Figure 7. PSNR of the watermarked MPEG-4 "Foreman" sequence (left) and of the MPEG-2 "Miss America" sequence (right) at different compression ratio versus the BER of a simulated communication link.

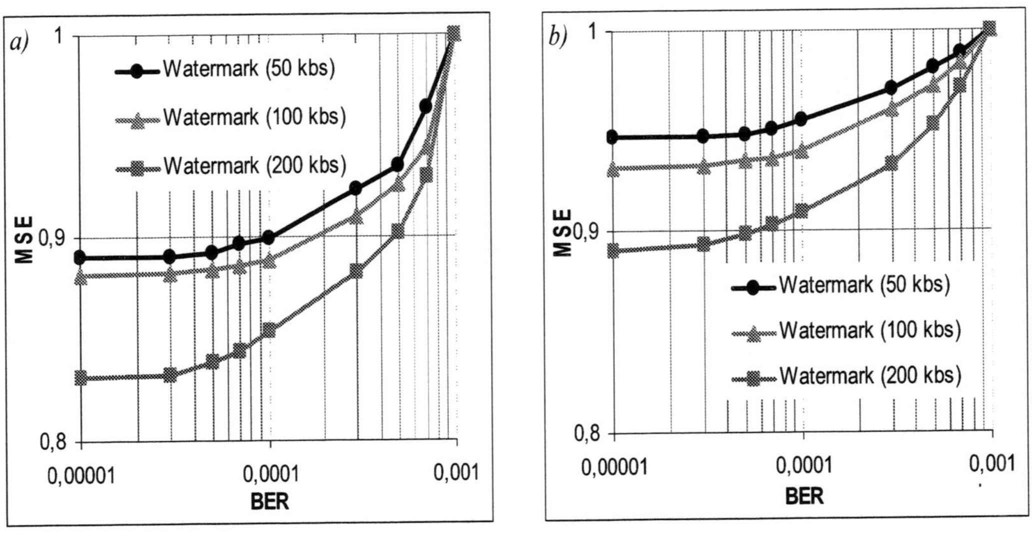

Figure 8. MSE (normalized to 1) of the estimated watermark with respect to the original one at different compression ratio for the MPEG-4 coded video sequences: *a)* "*Carphone*"; *b)* "*News*".

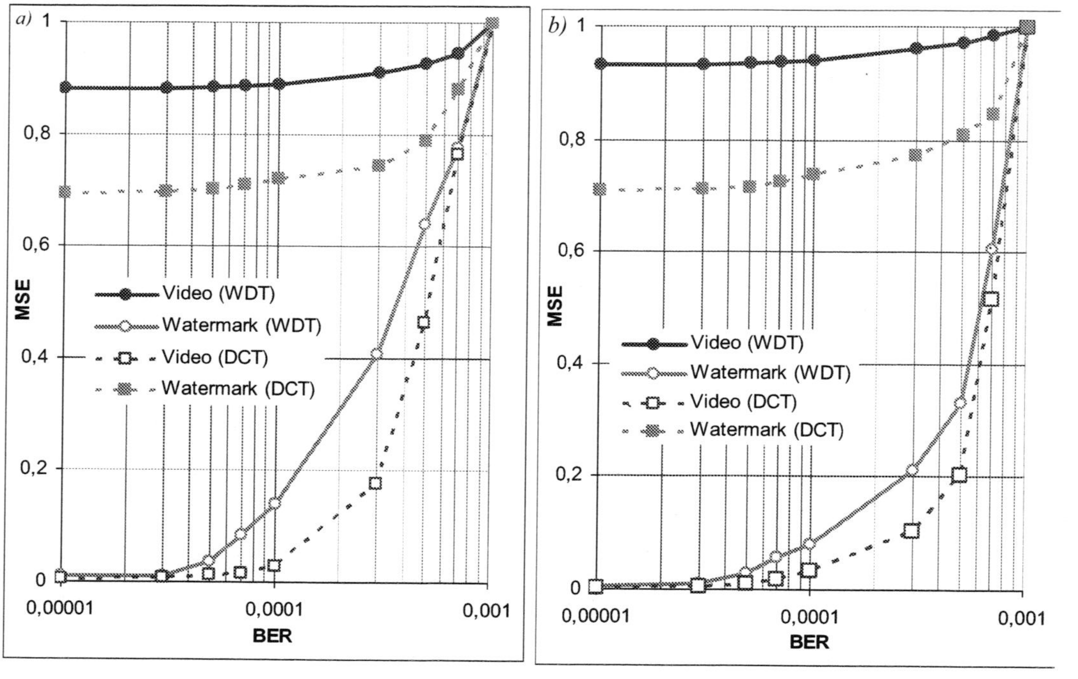

Figure 9. Watermark MSE and video sequence MSE (normalized to 1) versus the BER: *a)* for the sequence "*Carphone*" MPEG-4 coded at 50 Kb/s; *b)* for the sequence "*News*" MPEG-4 coded at 50 Kb/s (right).

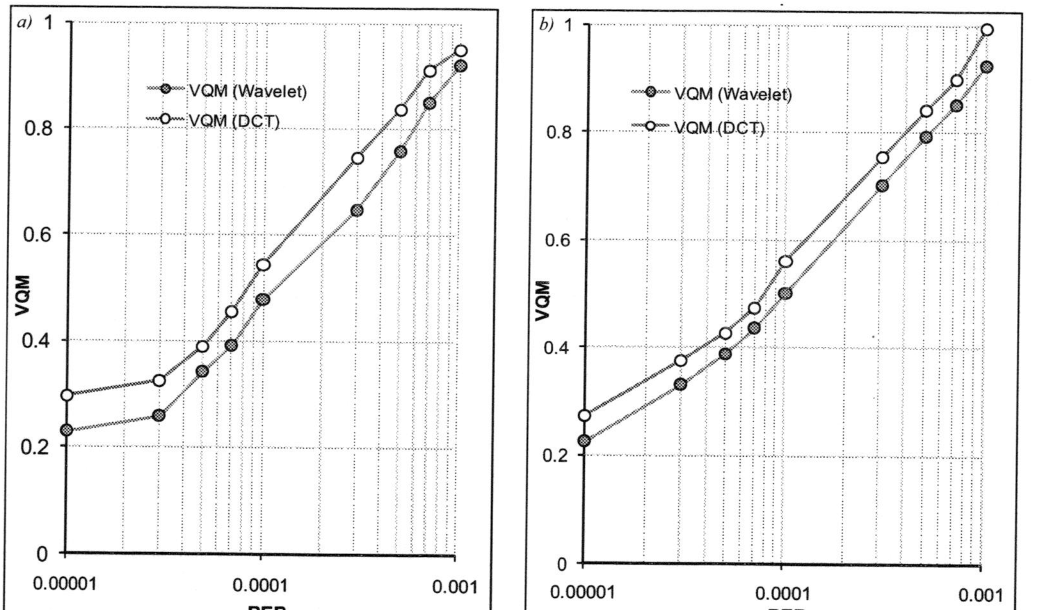

Figure 10. VQM of the video-sequence versus the BER of the simulated communication link for: *a)* the watermarked MPEG-4 "*Foreman*" sequence; *b)* the watermarked MPEG-4 "*News*" sequence.

V. Importance of Color in Video Conferencing Application

The efficient digital representation of image and video signals has been subject of considerable research over the last years, but only recently consumer electronics have increased demand for efficient and reliable video communications [29]. Video consumer devices, such as digital television (DTV), mobile video terminals and video-telephony have been quickly developed and visual communication has become more feasible than ever due to the growing availability of digital transmission links, progress in signal processing and image compression research. Digital image and video coding techniques develop from a more or less purely academic research area into a highly commercial business [30]. No longer is the conventional voice-call over a wired network the only reasonable and reliable method of transmitting information. Instead, the options are many and diverse, ranging from video-calls over the conventional wired network to video-telephony and video-calls over wireless/mobile networks [31].

In typical video-calls, each customer transmits both voice and MPEG-4 video images (usually his/her face in the foreground). This means that the face of the user is commonly the most important region of interest in video-phone applications, being the colour video-phone image composed by natural *head-and-shoulders* scene [32], [33]. Moreover, the emergence of the MPEG-4 standard [34] has made possible new ways to compactly represent audio-visual information. Specifically, the object based representation in the MPEG-4 standard permits the encoding transmission, and decoding of different objects in a selective fashion [35]. In the case of video-telephony, the object-based representation means that the foreground (i.e. a person facing the camera) and background objects in the scene can be encoded in such a fashion that allows the foreground to be decoded independently of the background. In particular, since in 3G video-phone communications the foreground is clearly more important than the background, the quality of the foreground object can be enhanced compared to the background. The broad goal of this communications revolution is to provide seamless and high-quality communications between people (or groups of people), *any-where*, *any-time*, *any-way*, and at a *reasonable* price [36].

Addressing some of these issues, we now focus on typical video-calls in 3G communication scenarios proposing a novel colour space to minimize the perceptual distortions on face coding introduced by tracing watermarking. More specifically, faces are *essential* for visual communication and could therefore be coded at a higher quality [37]. Moreover, many studies have shown that colour is very relevant in face and skin analysis. Indeed, skin is characterized more by chromatic than by luminance component [40]. This concept was also suggested in [41], where the authors investigate the importance of the skin colour for application to still images. We introduce a novel colour space, based on the "*skin*" component, in order to minimize the perceptual distortions introduced by digital watermarking. We have defined a new colour space, called *YST*, in which the luminance component is the same of *YUV*. Conversely, the *S* (skin) component has been obtained estimating the average values corresponding to a set of different colours of human faces. The criterion for this selection was to achieve a reasonable degree of generalization as it will be detailed in the following of this paper. The third component, *T*, is defined as orthogonal to the *YS* plane. Analytical results are used to show the benefits obtained in digital watermarking by this new representation. In fact, the sensitivity of the *YST* representation outperforms the conventional one in terms of both objective (mean square error, peak signal-to-noise ratio) and subjective (video quality metrics) indicators (see the following Sections).

VI. Color Spaces

Many studies have shown that colour representation is very relevant in image processing and communications for face and skin analysis. Indeed, skin is characterized more by chromatic than by luminance components [40]. Even for different people, chromatic skin features are relatively constant while illumination can vary a lot. The main idea in usual methods is to transform *RGB* signal to make luminance explicit so that it can be discarded. Only chromatic information is kept and used for the adopted image processing technique. Skin colour detection is a very important step in many vision systems, like gesture recognition, hand tracking, video indexing, and face detection [42]. Colour is the way the human visual system measures a part of the electromagnetic spectrum, approximately between 300 and 830 nm [43]. Because of certain properties of the human visual system, we are not able to see all of the possible combinations of the visible spectrum hut we tend to group various spectra into colours. The human skin colour occupies a certain range of luminance values in a colour space [44]. A Colour space is a notation by which we can specify colours, i.e. the human perception of the visible electromagnetic spectrum [43]. Colour spaces are, hence, different representation of colours using several methods. Examples include *RGB*, *YCrCb*, *YUV*, etc.

The choice of the colour space can be a very important decision which can dramatically influence the results of the processing. The knowledge of various colour spaces can ease the choice of the appropriate colour space. The *RGB* colour space is good for image display but is not the best when analyzing images using the computer. In fact, one of the problems of the *RGB* colour space is its perceptual non-uniformity, i.e. its low correlation between the perceived difference of two colours and the Euclidian distance in the *RGB* space. Moreover, the main disadvantage of the *RGB* colour space in applications with natural images is a high correlation between its components: a value of about 0.78 for the cross-correlation between the *B* and *R* channel, a value of 0.98 and 0.94, respectively, for the correlation between *R* and *G* and between *G* and *B* [43]. Because of this high correlation between the channels, the *RGB* domain is, hence, not suitable for image processing techniques, such as digital watermarking applications [45]. The potential of these three channels can be exploited for the application of watermarking, by decreasing the correlation among them. Other colours systems exist which have the property of separating the luminance component from chromatic component and with that at least partial independence of chromaticity and luminance is achieved. Such colour spaces are for example *YCrCb* and *YUV*, where *Y* stands for "*luminance*" and represents the luminance, while *Cr*, *Cb*, *U* and *V* represent the chrominance components, providing colour information and are "colour difference" signals of blue minus *luminance* (*B-Y*) and red minus *luminance* (*R-Y*), respectively. In practices, an image can be presented in a number of different colour space models such as [46]:

- *RGB* (as said before) stands for the three primary colours: red, green and blue. It is a hardware-oriented model and is well known for its colour-monitor display purpose.

- *YUV* is another hardware-oriented model. However, unlike the *RGB* space, here the luminance is separated from the chrominance data.

- *HSV* is an acronym for hue-saturation-value. *Hue* is a colour attribute that describes a pure colour, while *saturation* defines the relative purity or the amount of white light mixed with a *hue*; *value* refers to the luminance of the image. This model is commonly use for image analysis but it is not suitable for video coding.

These are some, but certainly not all, of the colour space models available in image processing. Therefore, it is important to choose the appropriate colour space for modelling human skin colour. Moreover, these colour spaces are *deformations* of the *RGB* colour space, usually obtained as linear transformations of the *RGB* space through a process called colour space conversion (CSC) [43]. Following this methodological approach, in the next Section we propose a novel colour space, *YST*, to model the human skin in order to minimize the perceptual distortions introduced on the skin colour component by the digital watermarking procedure previously described.

VII. The New Color Space "YST"

In the first half of this Section, we illustrate the constraints the new colour space *YST* must satisfy and how to obtain the *skin* component. In the second half, we obtain the third vector *T* with the constraint that it is orthogonal to the plane spanned by *Y* and *S*.

The new colour space must satisfy the following conditions: the luminance must be the same of the *YUV* colour space because *Y* is the component used in MPEG-4 for motion-compensation and *must* remain unaltered; one component, *S*, must be *ad hoc* created to match the vector corresponding to the skin colour and finally, the conversion *to* and *from* this new colour space must be reversible. In the *YST* colour space, the vector corresponding to *S* has been obtained estimating the mean values representing the skin colour (i.e. the only *U* and *V* components) of human faces, obtained from a cross-matching between a set of simulations, made on test images and video-sequences accounting for a randomized set of different skin colours, as well as works already published in the literature [44], [45], [47]. We have found the parameters of the vector S in full agreement with the methodological approach of [49], where the three preferred skin colour ellipses according to the ethnic group of Mongoloid, Caucasoid, and Negroid were determined. We have then modelled the "average" skin colour vector S according to the ethnic groups in the (U, V) chromaticity coordinates like done in [50]. The criterion for this selection was to achieve a reasonable degree of generalization. As said before, the values obtained in this way for the vector skin stand for the average colour of a human face. We have experimentally found that the vector *skin* is hence defined in the *YUV* domain by the following vector:

$$S = [0, 0.7072, 0.6345] \tag{4}$$

The vector *S* lays in the plane identified by the chrominance components *U* and *V*, i.e. it is a linear combination of only these two components. In this way, the vector *S* represents the mean value of the human "skin", in the sense that it stands for the *average chrominance* component of the human skin. It has to be noted that the vector S is characterized by unitary modulus (i.e. S is a *versor*). This means that S represents the *direction* of the average chrominance component of the human skin. In other words, the vector S corresponds to infinite human skin with different tones of luminance given by the combination of S with the *luminance* Y. This means that the chrominance of the skin remains nearly the same for different kind of people (i.e. different genders) while the luminance changes, from white to black, corresponding to different linear combinations of S and Y.

Once we have two of the three components, the third one, the vector T, is automatically identified in order to have a component that is orthogonal to the plane spanned by the other two vectors Y and S. Hence T can be used as the component for any image processing technique of face coding, e.g. watermarking of video-phone calls, in order to minimize the perceptual distortions introduced in the image by the processing. In order to identify the third component without any degree of freedom, we must impose another condition: the range of T must be the same of the luminance Y. This latter condition follows directly from the fact that we want to use the vector T for any image processing applications, showing the benefits of our approach in respect to other works already published in the literature [9], where the watermark embedding process was implemented using the vector Y of the *YUV* colour domain. With respect to the previous conditions, the value of the third component T can be easily derived; we have verified that it is defined, in the *YUV* domain, by the following vector:

$$T = [0, -0.6345, 0.7072] \quad (5)$$

In practice, we have generated the new colour space *YST* in which the third component, the vector T, is the component carrying less information about the original image. The transformation matrix to obtain the conversion from the *YUV* domain to the new *YST* colour space is then given by:

$$\begin{vmatrix} Y \\ S \\ T \end{vmatrix} = \begin{vmatrix} 1 & 0 & 0 \\ 0 & 0.7072 & 0.6345 \\ 0 & -0.6345 & 0.7072 \end{vmatrix} \cdot \begin{vmatrix} Y \\ U \\ V \end{vmatrix} \quad (6)$$

where it clearly appears that the luminance component is the same for both the two colour domains. It is interesting to point out that in the novel colour space *YST*, as well as in *YUV*, the luminance component is decoupled from the colour information of the image. As a consequence, the skin-colour model can remain effective regardless of the variation of skin colour (e.g., black, white, or yellow) because the derivation of the model is independent of the luminance information of the image, as already stated in [46]. Moreover, the conversion from and to the *YUV* space colour is reversible and identified, respectively, by equation (6) and by its inverse:

$$\begin{vmatrix} Y \\ U \\ V \end{vmatrix} = \begin{vmatrix} 1 & 0 & 0 \\ 0 & 0.7834 & -0.7029 \\ 0 & 0.7029 & 0.7834 \end{vmatrix} \cdot \begin{vmatrix} Y \\ S \\ T \end{vmatrix} \quad (7)$$

The new colour space *YST* has been obtained from the *YUV* domain with a linear transformation and, hence, it can be easily related to all the other colour spaces that can be derived, providing calibration data, from (and to) the *YUV* representation (e.g. *RGB*, *YCrCb*, *HSV*, etc…).

VII). Results and Discussion

This section shows the benefits obtained by the new colour space representation *YST* using a tracing watermarking as the adopted image processing technique for QoS assessment. In order to minimize the perceptual degradations introduced by digital watermarking, the mark will be inserted in the third component *T*. In the following we have used both objective and subjective indicators to evaluate the performance of the embedding process. Analytical results are used to show the benefits obtained by the new colour space representation as well as in terms of mean square error (MSE), peak signal-to-noise-ratio (PSNR), and video quality metric (VQM). It has to be noted that the aforementioned quality metrics, corresponding to the two different colour spaces, can be easily compared since both the vectors *S* and *T* have unitary modulus.

The dimensions of the video sequences employed in our experimentations have been properly chosen in order to simulate a multimedia service in a UMTS scenario, i.e. typical 3G video-calls. Therefore, QCIF (144x176) video sequences, which well match the limited dimensions of a mobile terminal's display, have been employed. The marked video is then transmitted over a noisy channel that is simulated by a Poisson's generator of random transmission errors. The QoS assessment procedure has been tested for both the MPEG-2 and the MPEG-4 coders. Different bit rates (600, 300, 100, and 50 Kb/s) have been considered in our experiments. We have first verified that using the new colour space is still possible to *trace* the alteration suffered by the data in the transmission channel. Then, we have evaluated the benefits of using this new colour representation (i.e. watermark in *T*) instead of embedding the watermark in *Y*, as illustrated in [9].

Figure 11 shows the PSNR and the VQM of one watermarked frame (both in the *YUV* and *YST* domain) of the video-sequences used in our analysis, versus the watermark intensity (i.e. varying the value of the parameter β): Figures 11 (*a, b*) refer here to the first frame of the video-sequence *Miss America*, Figures 11 (*c, d*) to the video sequence *Akiyo* while Figures 11 (*e, f*) refer to the *News* sequence. According to the VQM indicator, it has to be noted that the curves related to the new embedding domain (the novel colour space) are always lower than the same curves obtained with the conventional approach. This means that the perceptual degradations introduced by the watermark embedding process in the image have been minimized. Moreover, the quality of the image decreases both in terms of objective (PSNR) and subjective (VQM) indicators increasing the watermark intensity.

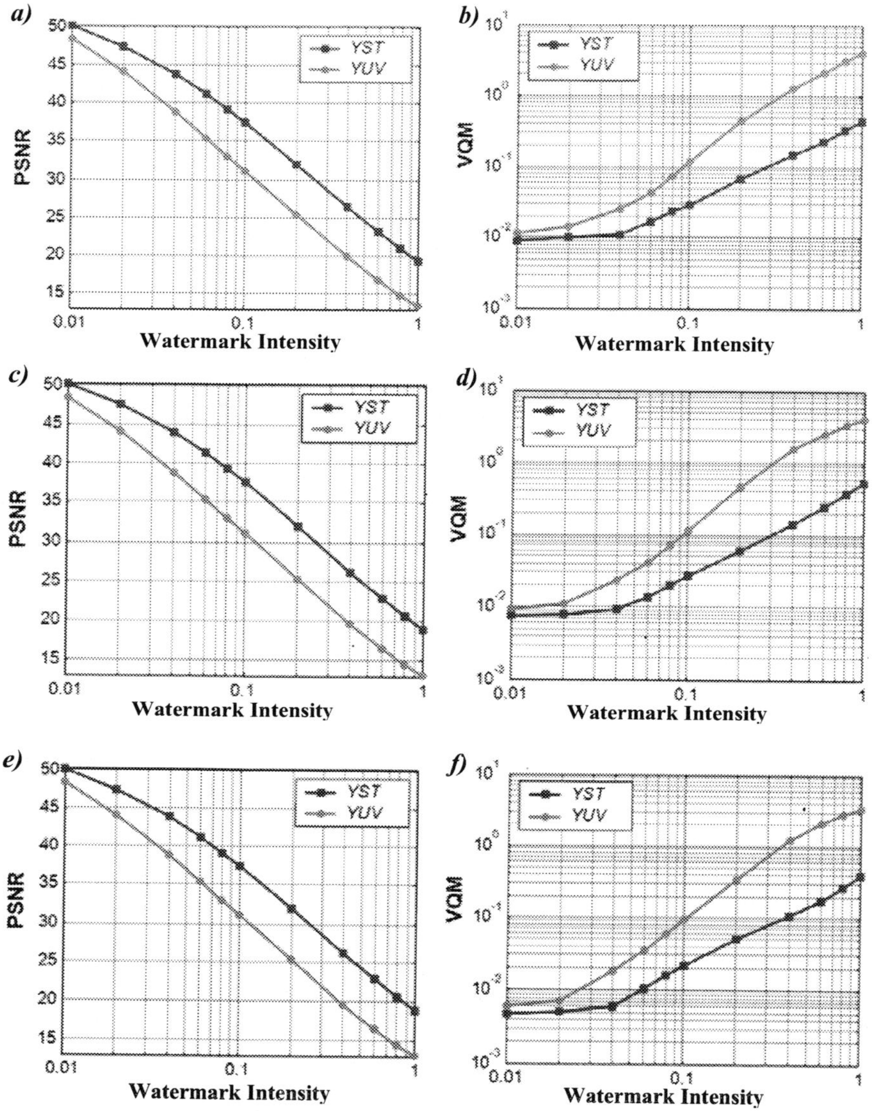

Figure 11. PSNR (left) and VQM (right) versus the watermark intensity of the first watermarked frame for the video-sequence: *Miss America* (a, b); *Akiyo* (c, d) and *News* (e, f)

For a visual comparison, in order to evaluate the assessment of the visual quality of the image after the embedding process, we have put side by side one frame of the original video-sequence versus the same frame after the embedding process (realized in the *YUV* and in the *YST* colour domain). In particular, in Figures 12 (*a*, *b*, and *c*) we show one frame of the original video-sequence (*Miss America*, *Akiyo*, and *News*, respectively) without compression (*upper left*) and after the MPEG-4 coding (*upper right*) compared with the same frame watermarked in the *YUV* colour space (*lower left*) and in the *YST* domain (*lower right*). The three video sequences from which the four frames in Figures 12 (*a-c*) (namely, original, MPEG-4 compressed, MPEG-4 compressed after the luminance *Y* watermarking, MPEG-4 compressed after the watermarking in the *T* component) can be visually compared after downloading the zipped data folder from the WEB site [48], password *color*. The embedding process has been here realized with a watermark intensity $\beta = 0.4$ (much stronger than in [9]), in order to make the perceptual degradations in the frames clearly visible. Using a stronger watermark in the new embedding domain, but with the same perceptual degradation of the conventional approach, reflects in a more effective watermark extraction process. The benefits of such a choice are twofold: first, we can minimize the perceptual degradations introduced by the embedding process, using the same watermark intensity; second, we enhance the dynamic properties of the QoS evaluation system, because we need a smaller number of frames in order to extract the watermark from the video-sequence, drastically reducing the time requested by the quality assessment procedure to be evaluated.

(a) (b)

(c)

Figure 12a. Visual assessment of the first frame of the video-sequence: *Miss America*: original frame without compression (*upper left*), original frame after MPEG-4 coding (*upper right*), MPEG-4 coded frame watermarked in the *YUV* domain (*lower left*), MPEG-4 coded frame watermarked in the *YST* domain (*lower right*). The watermark intensity has been posed equal to $\beta = 0.4$ in order to make the perceptual degradations in the images clearly visible.

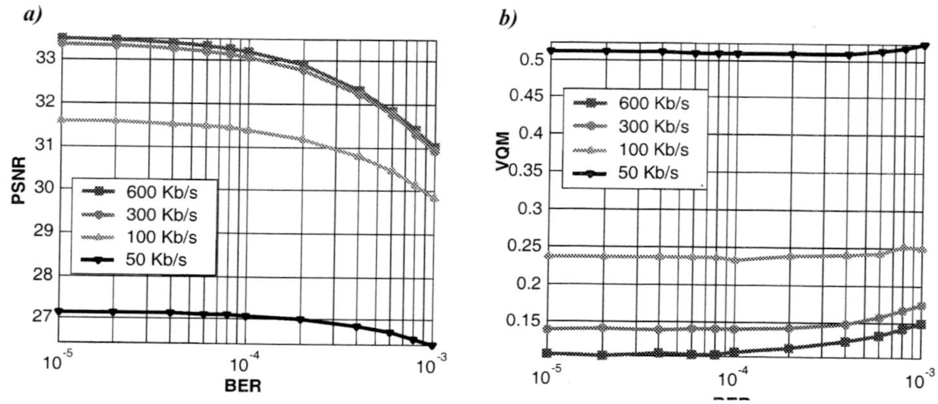

Figure 13. PSNR (left) and VQM (right) of the watermarked MPEG-2 "*Akiyo*" sequence at different compression ratio versus the BER of a simulated communication link.

Conversely, the weaker watermark power factor $\beta = 0.04$ has been used in all the following simulations to meet the watermark settings of [9] and compare the results to the ones presented in that paper. In order to characterize the performance of the proposed method to provide a quality measure of the coding/transmission process of the video, Figures 13 and 14 show both the PSNR and VQM of the received video versus different BER of the transmission channel. The procedure has been tested for the MPEG-2 coder (see Figure 13) as well as for the MPEG-4 coder (see Figure 14) and different compression ratios have been considered. It is worth nothing that the PSNR (Figures 13*a*, 14*a*) and the VQM (Figures 13*b*, 14*b*) of the video-sequences decreases when the BER increases and the bit-rate decreases. This is in accordance with the perceptual degradation that the videos suffer at increasing BER and decreasing bit-rate. In addition, it has to be underlined that the quality degradation of the extracted watermark embedded into the host video has the same behaviour of the one affecting the video itself, as shown in terms of MSE in Figure 15.

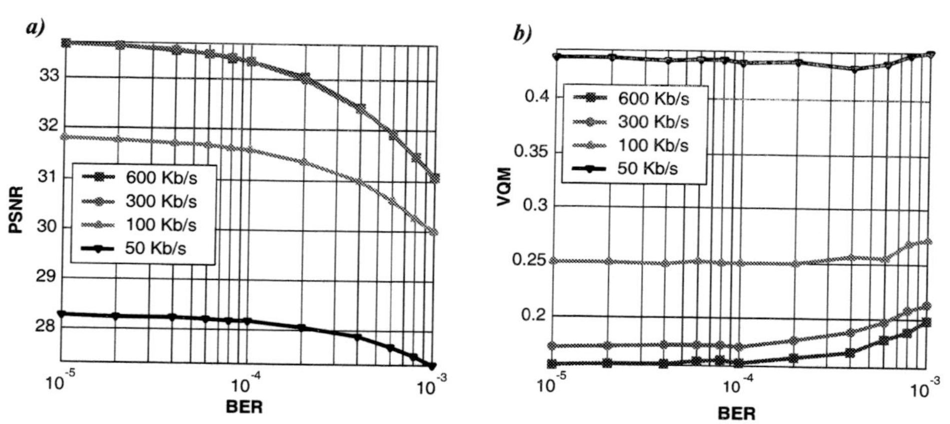

Figure 14. PSNR (left) and VQM (right) of the watermarked MPEG-4 "*Miss America*" sequence at different compression ratio versus the BER of a simulated communication link.

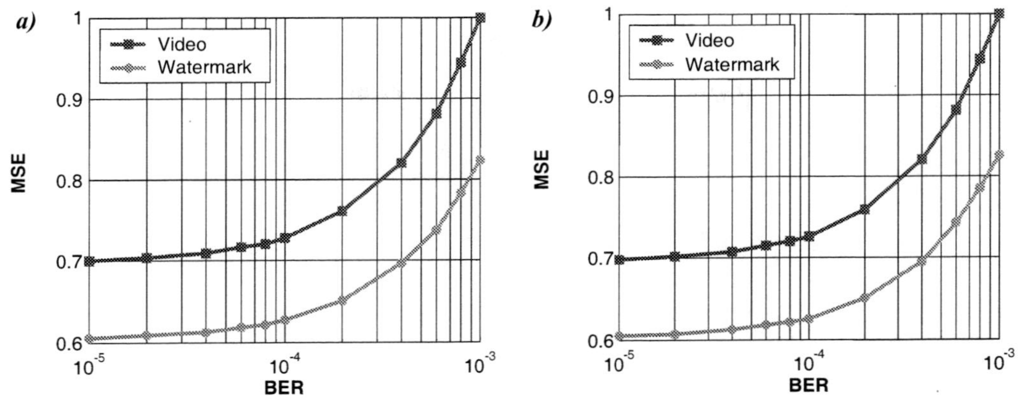

Figure 15. Watermark MSE and video sequence MSE (normalized to 1) versus the BER: for the MPEG-2 "*Akiyo*" sequence coded at 100 kbit/s (left) and for the MPEG-4 "*Miss America*" sequence coded at 100 kbit/s (right).

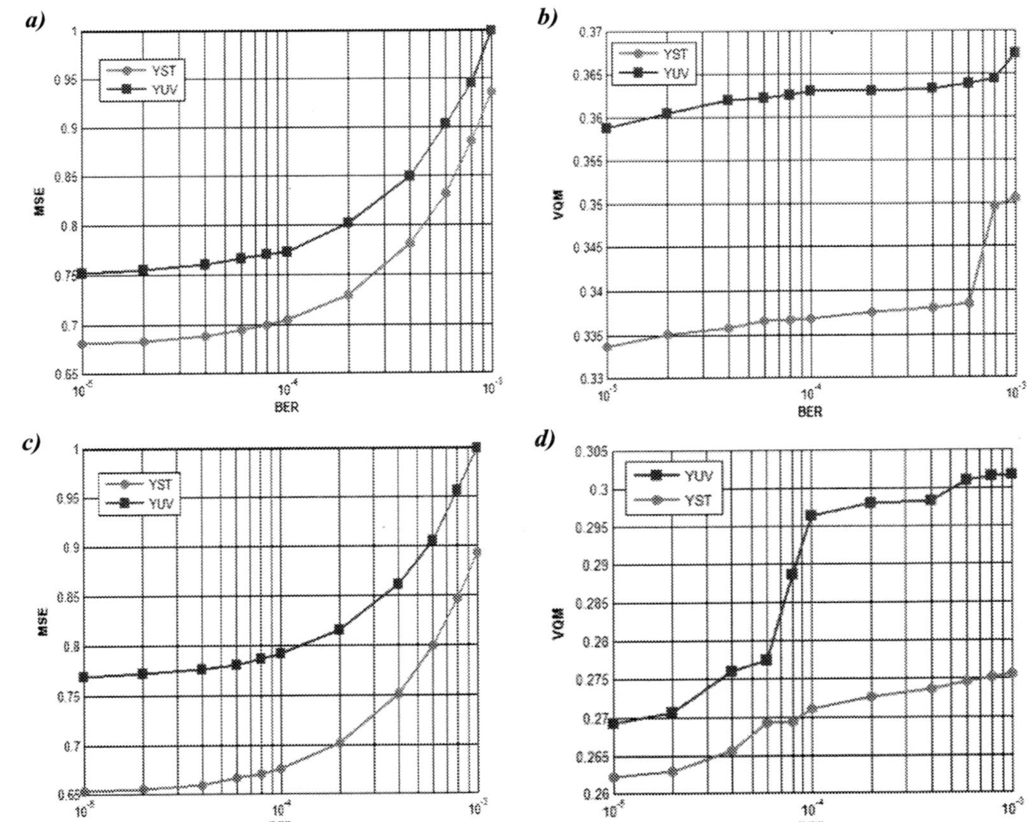

Figure 16. MSE of the extracted watermark (*left*) and VQM of the video-sequence (*right*) versus the BER of the simulated communication link for: the watermarked MPEG-4 "*Foreman*" sequence (*a, b*) and the watermarked MPEG-2 "*News*" sequence (*c, d*) both coded at 100 kbit/s after watermark insertion in the component *T* for the *YST* domain, and in the component *Y* for the *YUV* representation.

Moreover, in order to quantify the benefits of the new colour representation, we have compared our simulation trials with the results obtained using the *YUV* colour domain. As a consequence, the embedding procedure has been realized inserting the mark in *Y*, for the *YUV* representation, and in *T* for the new approach, varying the BER of the simulated channel. Figs 16 (*a*, *b*) relate to the "*Foreman*" sequence MPEG-4 coded at 100 kbit/s while Figures 16 (*c*, *d*) refer to the "*Miss America*" sequence MPEG-2 coded at 100 kbit/s. In particular, Figure 16*a* and Figure 16*c* show the MSE of the extracted watermark after the coding/transmission process while Figure 16*b* and Figure 16*d* evidence the VQM of the two video-sequences, versus the BER. As it can be easily seen from these figures, promising results has been obtained, using both objective and subjective indicators for both MPEG-2 and MPEG-4 coders.

It is worth pointing out that the mobile station (MS), that is the end-user of the communication process, needs to evaluate the actual QoS provided by the network operator (see the conceptual scheme in Figure 3). As a consequence, the MS must perform real-time processing to implement some kind of evaluation of the supplied QoS.

Let us now focus on the computational complexity of the proposed technique, by analysing the number of the operations performed by the MS for the QoS assessment. According to the scheme in Figure 5, the floating point operations (flops) to be implemented by the MS processor are: first, the change of the chromaticity coordinates of every frame (i.e. from the colour space YUV to the new colour space YST); second, the computation of the K_1-by-K_2 DCT of each frame; third, the correlation with the spread version of the watermark made of $\frac{1}{4} \cdot K_1 \cdot K_2$ pixels (i.e. the original watermark pre-multiplied by one known PN matrix, different frame by frame); fourth, the accumulation of the detected mark averaged over all the M frames; last, the final evaluation of the MSE of the detected mark at the end of the procedure. The whole computation can be implemented either by sequential or parallel processing, depending on the available hardware. The approximate computing time needed by the QoS assessment procedure (from M frames of K_1 by K_2 pixels), operating either on sequential or parallel processors on a mobile device, is depicted in Table 1 in terms of the time required by the number of flops. For sake of compactness, we have reported only the number of real products and sums, assuming that:

Table 1. Approximate number of the flops-time required for the QoS detection on a mobile device

	Sequential processing		Parallel processing	
	No. of real products	No. of real sums	No. of real products	No. of real sums
Chromaticity coordinates	$2\,M \cdot K_1 \cdot K_2$	$M \cdot K_1 \cdot K_2$	M	M
DCT	$8\,M \cdot K_1 \cdot K_2 \cdot \log_2(4\,K_1 \cdot K_2)$	$8\,M \cdot K_1 \cdot K_2 \cdot \log_2(4\,K_1 \times K_2)$	$M\,[\log_2(2\,K_1) + \log_2(2\,K_2)]$	$2\,M\,[\log_2(2\,K_1) + \log_2(2\,K_2)]$
Correlator	$\frac{1}{4}\,M \cdot K_1 \cdot K_2$	0	M	0
Accumulator	0	$\frac{1}{4}\,M \cdot K_1 \cdot K_2$	0	M
MSE	$K_1 \cdot K_2$	$2\,K_1 \cdot K_2$	1	$\log_2(K_1 \cdot K_2)$

- in sequential implementation, the time of one complex product is the same as four real products and two real sums, while the time of one complex sum time is equivalent to two real sums;
- in parallel implementation, the time of one complex product is the same as one real product and one real sum, while the time of one complex sum time is equivalent to one real sum.

In particular, sequential implementation needs 764.6 Mflops, while parallel implementation only requires 1.5 Kflops, to process one second of a 25-frames-per-second video-sequence in the CIF (352 x 288) format. Using the QCIF (176 x 144) format, 170.9 Mflops are necessary in sequential implementation while 1.4 Kflops are needed in parallel implementation. It has to be noted that in all the aforementioned cases, the DCT computation is the most time consuming step: 98.8 % of operations for sequential implementation and 92.2 % for parallel implementation are required for the only DCT processing over the whole computation time. The tremendous advantage of parallel computing is motivated by the fact that DCT computation is implemented by Fast Fourier Transform (FFT). In fact, the computational complexity of FFT (over N data) is of the order of N $\log_2 N$ while it reduces to the order of $\log_2 N$ by parallel processing schemes.

It is opinion of the authors that the implementation of the proposed technique is going to be feasible on a mobile device since the MS, in a short time, will host large processing capabilities (including hardware DCT co-processors) because of the monotonically decreasing cost of very large scale integration as well as the ever increasing running speed. Therefore, the complexity of the QoS evaluation procedure, which is presented in this paper, appears negligible in comparison with real time MPEG decoding and adaptive on-board array processing."

The obtained results evidence the sensitivity of the watermarking quality index to the actual quality for given target quality levels and show the capability of this *unconventional* use of watermarking techniques to trace the alterations suffered by the data through the communication channel. Moreover, our simulation trials have evidence the benefits of the new colour representation for digital watermarking application instead of using *YUV* as the embedding domain. We have verified that realizing the watermark embedding in the new colour space *YST* (specifically, inserting the mark in *T*) minimize the degradations suffered by the video during the processing as well as in terms of both objective and subjective indicators.

In particular, such QoS indexes can be usefully employed for a number of different purposes in wireless multimedia communication networks such as: control feedback to the sending user on the effective quality of the link; detailed information to the operator for billing purposes and diagnostic information to the operator about the communication link status.

From our wide simulation trials, we have experimentally verified that the QoS measurement (i.e. the watermark extraction) is able to operate every two seconds (or even less) with a corresponding detection probability of 95% (at least). The QoS index can be sent back to the network manager in order to adopt optimal policies in terms of QoS adjustment to the subscriber's duplex link. In particular, the QoS index could bring the UMTS network operator to force a (likely soft or softer) handover process, in the cases when the power control fails. Handover procedure is usually performed to serve the customer by another cell characterised by less interference or stronger pilot signal reception. In fact, user's mobility is the main reason for handover, usually established in UMTS by the inner loop fast power control according to the quality target of the outer loop control. But system load control or "bad" (multipath) channel conditions may also cause one further radio link addition (branch addition). The QoS index could act as a supplementary warning when the power control is unable to request a handover for that subscriber. In practice, the mobile's quality index may be fed back to the network manager to eventually force the handover process if the power control has been ineffective. In this way, the video-call quality can be restored in a short time and, additionally, the operator can more accurately fare the subscriber for the supplied QoS.

IX. Dynamic QoS Management in 3G Communications: "Boomerang" Watermarking

For applications running in a shared environment, the allocation and management of these resources is an important question, although most existing systems are based on a best-effort approach [38]-[39]. Due to their real-time characteristics, multimedia sessions need quality guarantees from the network [51]. In addition, for a teleconferencing application involving many users, a single quality of service level may not be appropriate for all participating users, since some users may participate with a very limited local workstation which cannot provide for the quality which is adopted by the majority of the conference participants. Consequently, multimedia content processing has employed new paradigms that include distributed, cooperative, and resource-constrained models, whereas multimedia communications have addressed the new application challenges by focusing on cross-layer design [52]. With few exceptions, such as joint source channel coding and adaptive streaming, processing of multimedia content has been approached as a network-independent problem by the research and academic communities [53]. In turn, the works that addressed the content delivery aspects have typically not included the characteristics of the source content and have primarily studied interactions among lower layers of the protocol stack [54]-[55]. However, the processing and delivery of multimedia content are not independent and their interaction has a major impact on the QoS aspects. QoS is designed to hide low-level application variation and provide necessary service guarantees.

In this Section, a network-aware processing of multimedia content is described (double boomerang approach) for QoS management exploiting the tracing watermarking procedure previously described. Each user produces a video-sequence and this multimedia content is transmitted to the receiver by means of the communication channel. The basic idea of using a two-way spread spectrum communication (called double boomerang transmission system) was exploited in [55] for the purpose of sharing data and positioning information in vehicular networks.

The received video sequence, once decoded, is split in two ways, one to the user's display, the other to employ the watermarking procedure. Thus, the same information travels once to the receiver, in a transparent mode, and then to the transmitter, in a hidden mode, describing a boomerang trajectory. In addition to this, the system is collaborative, that is to say every user employs the received information to watermark his video-sequence before the transmission. Watermark embedding is led from an opportune transformation of the received frames: every 2-3 seconds a designed frame is extracted from the video-sequence, its central area is cropped, and then binarized to decrease the information bits to be embedded, and finally used as watermark. This is a consequence of the fact that, in typical video-calls, each customer transmits both voice and MPEG-4 video images (usually his/her face in the foreground). This means that the face of the user is commonly the most important region of interest in video-phone applications, being the colour video-phone image composed by natural head-and-shoulder scene. Hence, each customer uses as watermark the central part of the video-sequence received from the other user: each user receives a watermarked video-sequence, where the watermark embedding procedure is the same of the one described in Section III but the watermark changes dynamically. When this watermarked sequence is received at the other side of the communication link, the receiver knows the hidden information that contains because the receiver is the owner of the multimedia content (i.e. the watermark is a frame of the sequence previously transmitted). By extracting the watermark from the sequence, the agent can compare this watermark with the original one (i.e. the high quality version of the watermark).

This mechanism allows a dynamic control of network awareness and, according to previous studies [9], the watermark suffers the same degradation as the multimedia hosting content. As a consequence, using the MSE metric, each user can estimate the QoS in the communication link, comparing the received watermark with its original counterpart stored in the user equipment. The collaborative processing previously described allows detecting the problems of the communication link because the tracing watermark model is employed between both the two active users involved in the videoconference call. The sensitivity of such approach will be validated in the following through extensive simulation trials applied to typical operating scenarios of wireless/mobile communications.

Now, some experimental results characterizing the effectiveness of the proposed method are presented. The dimensions of the video-sequences employed in our experimentations have been properly chosen in order to simulate a multimedia service in a UMTS scenario, i.e. typical 3G video-calls. Therefore, QCIF (144x176) video sequences, which well match the limited dimensions of a mobile terminal's display, have been employed and a frame rate of 15 fps has been chosen. The marked video is then transmitted over noisy channels, simulated by Poisson's generators of random transmission errors. Specifically, wireless channels characterized by different levels of bit error rate (BER) have been designed, considering the customers near and/or far from the base station (BS), respectively. Specifically, we take into account the near-far-effect typical of cellular systems: when the user is near the BS, the uplink communication has been considered worse than the downlink path in terms of BER while, when the user is far from the BS, the uplink path has been considered better than the downlink (i.e. the mobile transmits with much more power to compensate the distance from the BS). For users closer to the BS, we have imposed an uplink BER of 10^{-4} and a downlink BER of 10^{-5}, while for users far from the BS, the uplink BER has been considered equal to 10^{-4} and the downlink BER to 10^{-3}. Our simulation trials are proposed to test the effectiveness of the double boomerang approach. Four situations are simulated: two mobile agents far from the BS (scenario 1), two mobile agents near to the BS (scenario 2), the first mobile agent is near while the second is far from the BS (scenario 3), the first mobile agent is far from the BS while the second is near (scenario 4). In all the following simulations, we have used the common test video-sequences such as *Carphone*, *Foreman*, *Miss America* and *Suzie*, all MPEG-4 coded. In particular, we have supposed two scenarios: a video-call between the users *Carphone* and *Foreman* and another video-call between *Miss America* and *Suzie*. We have first verified that with this double boomerang approach is still possible to trace the alterations endured by the data in the communication link. Then, we have evaluated an empirical relationship between the MSE of the watermark and of the multimedia content transmitted trough the network. Figure 17 (*a*) shows here the watermark MSE of the video-communication between *Miss America-Suzie*, while Figure 17 (*b*) depicts the watermark MSE of the *Carphone-Foreman* video-communication. The MSE metric is evaluated using (2), comparing the received watermark (embedded using the double-boomerang approach explained in Section 3) and its original counterpart, i.e. the original frame opportunely adapted. It has to be noted that the MSE behaviour is a direct consequence of the QoS of the network. In fact, the communication link is supposed to have the best quality in the first case (i.e. scenario 1), worst quality in the second case (i.e. scenario 2) and intermediate quality otherwise (i.e. scenarios 3 and 4). In other words, the degradations suffered by the watermark during the communication are proportional to the QoS (i.e. the BER) of the link: in fact, it can be easily seen from Figure 17 (*a*, *b*) that the watermark is highly corrupted in the second scenario (corresponding to the worst BER), while it is slightly damaged in the first scenario (corresponding to the best BER). In the third and fourth scenarios, the MSE is at an intermediate level because the BER of the channel has an intermediate value between the first and second scenario.

In order to characterize the performance of the proposed method to provide a quality measure of the transmission process of the multimedia content through the network, Figure 18 shows here the MSE of both the received video and the watermark, versus different BER of the transmission channel. To understand the relation between the multimedia content and the watermark, we compare the MSE of the watermark with the MSE of the video after the double boomerang trajectory. It has to be remembered that the video-sequence of every active user travels on the communication link one time as video (transparent mode), one time as watermark (hidden mode).

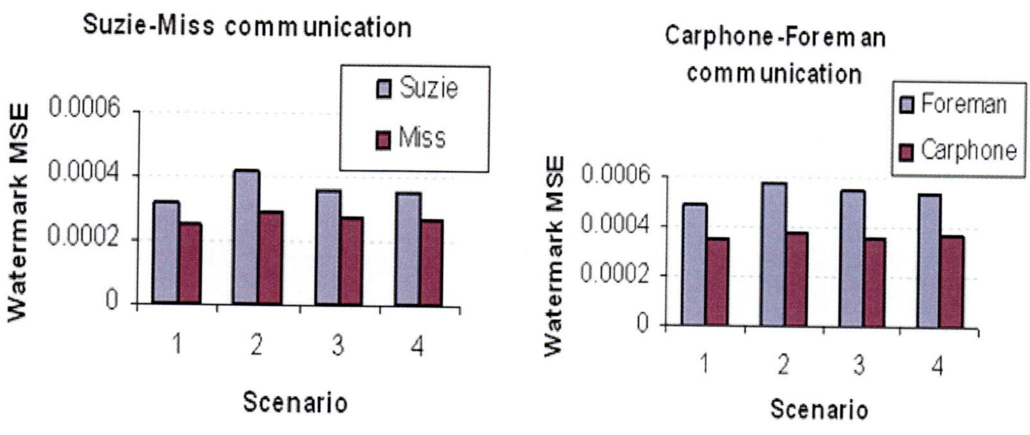

Figure 17. Watermark MSE versus different scenarios for the communication between: a) Suzie-Miss America; b) Carphone – Foreman.

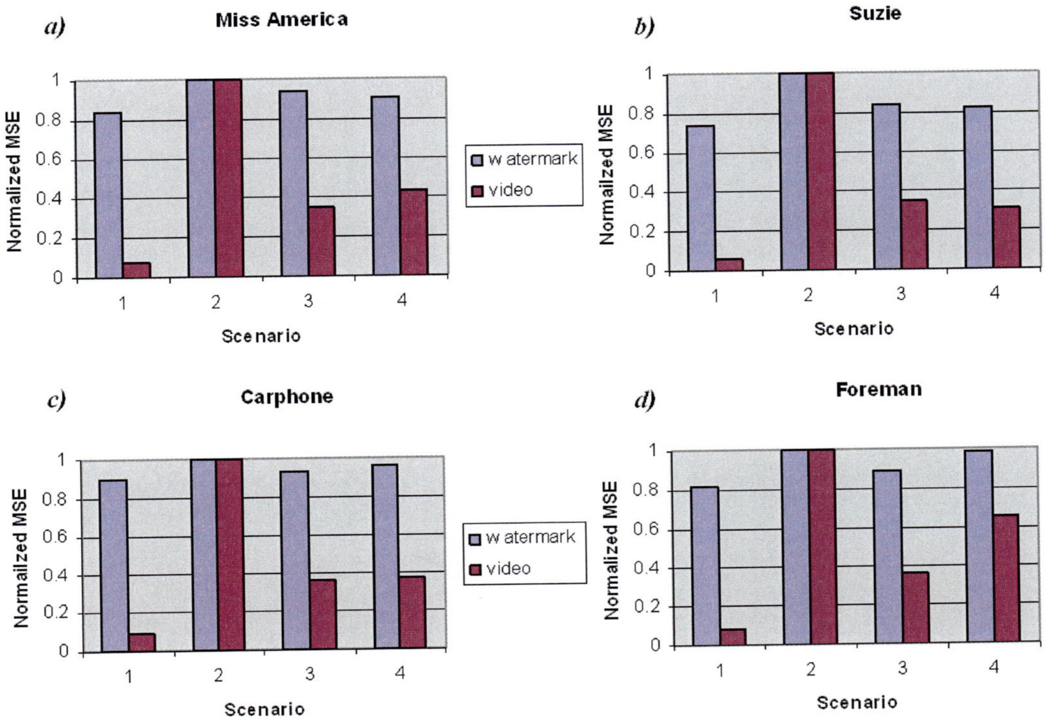

Figure 18. MSE (normalized to 1) versus different scenarios in the case of: *a)* Miss America; *b)* Suzie; *c)* Carphone; *d)* Foreman

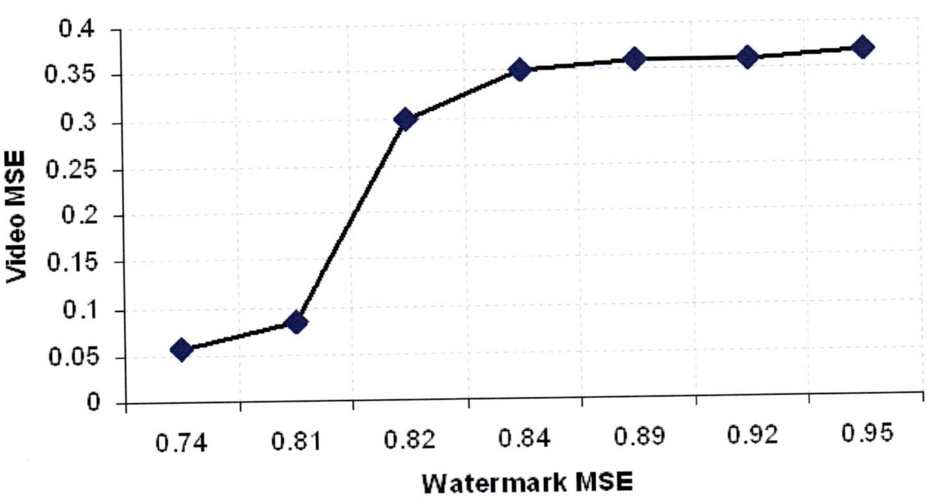

Figure 19. Relationship between watermark and video degradation in terms of mean square error.

Hence the watermark suffers twice the degradations of the communication link, due to the boomerang path. It is worth noting that the MSE of the video-sequences decreases when the BER increases: this is in accordance with the degradation that the videos suffer at increasing BER. In addition, the quality degradation of the extracted watermark embedded into the host video has the same behaviour of the one affecting the video itself, as shown in terms of MSE in Figure 18. Comparing these experimental results, it is now possible to evaluate an empirical relationship between the MSE of the boomerang watermarking and the one of the host multimedia content, as shown in Figure 19. The obtained results prove the sensitivity of the watermarking quality index to the actual quality for given target quality levels and show the capability of this network-aware processing of multimedia content to trace the alterations suffered by the data through the network.

X. Conclusion

This chapter discusses the impact (i.e. motivation and operating guidelines) of QoS assessment in video-conferencing applications. In particular, real-time interactive multimedia communications is investigated. We consider the end-to-end service quality methodologies presenting a QoS evaluation technique by means of an unconventional use of tracing watermarking. We first focus on embedding techniques in transformed domains (e.g. Fourier and Wavelet) and then, we concentrate on 3G video-conferencing communication scenarios, where each customer transmits both voice and MPEG-4 video sequences in real time. We propose and adopt a novel colour space (YST) to minimize the perceptual distortions on face coding introduced by the tracing watermarking, discussing the importance of the choice of the proper color space for image and video processing. Finally, we present a new watermarking method, for applications running in a shared environment, which allows a network-aware processing of multimedia content.

The advantages of distributing the management process can be summarized in: (i) an easier and more precise localization of the cause of QoS problems, (ii) better knowledge of local situations, (iii) a lower complexity for a single QoS agent and (iv) an increase in possible actions.

XI. References

[1] Chou, C. T. & Shin, K. G. (2004). "Analysis of Adaptive Bandwidth Allocation in Wireless Networks with Multilevel Degradable Quality of Service", *IEEE Trans. on Mobile Computing, vol. 3*, no. 1, 5-17, Jan.-Mar.

[2] Huang, V. & Zhuang, W. (2004). "QoS-Oriented Packet Scheduling for Wireless Multimedia CDMA Communications", *IEEE Trans. on Mobile Computing, vol. 3*, no. 1, 73-85, Jan.-Mar.

[3] Bader, F., Pinart, C., Cliristophi, C., Tsiakkouri, E., Ganchev, I., Friderikos, V., Bolloris, C., Correia, L. & Ferreira, L. (2003). "User-Centric Analysis of Perceived QoS in 4G IP Mobile/Wireless Networks", *14th IEEE Proc. on Personal, Indoor and Mobile Radio Comm., vol. 3*, 2047-2053, Sept.

[4] Giunta, G. (2003). 'Quality of service assessment in new generation wireless video communications', in: Digital Image Sequence Processing, *Compression, and Analysis*, chap. *5*, CRC press.

[5] Jiang, H. & Zhuang, W. (2003). "Quality-of-service provisioning to assured service in the wireless Internet", *IEEE Global Telecomm. Conf., vol. 6*, 3078-3082, Dec.

[6] Castro, J. P. (2001). 'Service Components in UMTS', in: The UMTS Network and Radio Access Technology: *Air Interface Techniques for Future Mobile Systems*, chap. *6*, J. Wiley & Sons.

[7] Lump, W. Y. & Lau F. C. M. (2003). "User-Centric Content Negotiation for Effective Adaptation Service in Mobile Computing", *IEEE Trans. on Software Eng., vol. 29*, no. 12, 1100-1111, Dec.

[8] Feng, N., Mau, S. C. & Mandayam, N. B. (2004). "Pricing and Power Control for Joint Network-Centric and User-Centric Radio Resource Management", *IEEE Trans. on Comm., vol. 52*, no. *9*, 1547-1557, Sept.

[9] Campisi, P., Carli, M., Giunta, G. & Neri, A. (2003). 'Blind Quality Assessment System for Multimedia Communication Using Tracing Watermarking', *IEEE Trans. Signal Proc., vol. 51*, no. 4, 996-1002.

[10] Campisi, P., Carli, M., Giunta, G. & Neri, A. (2002). 'Tracing watermarking for multimedia communication quality assessment', *IEEE Int. Conf. on Comm.*, New York.

[11] Campisi, P., Giunta, G. & Neri, A. (2002). 'Object based quality of service assessment for MPEG-4 videos using tracing watermarking', *IEEE Int. Conf. on Image Proc.*, Rochester, NY.

[12] Pakkala, D., Välitalo, P. & Latvakoski, J. (2003). "A User Centric Peer-to-Peer Service Environment for Interaction with Networked Appliances", in *Proc. of 23rd Int. Conf. on Distributed Computing Systems Workshops*, 242-247, May.

[13] O'Donnell, A. J. & Sethu, H. (2002). "A Novel, Practical Pricing Strategy for Congestion Control and Differentiated Services", *IEEE Int. Conf. on Comm., vol. 2*, 986-990.

[14] Ileri, O., Mau, S. C. & Mandayam, N. B. (2005). "Pricing for Enabling Forwarding in Self-Configuring Ad Hoc Networks", *IEEE J. on Sel. Areas in Comm., vol. 23*, no. 1, 151-162, Jan.

[15] Cox, I., Kilian, J., Leighton, F. & Shamoon, T. (1997). 'Secure Spread Spectrum Watermarking for Multimedia', *IEEE Trans. Image Proc., vol. 6*, no. 12, December.

[16] Hartung, H., Su, J. K. & Girod, B. (1999). 'Spread spectrum watermarking: malicious attacks and counterattacks' *Proc. SPIE, Security and Watermarking of Multimedia Contents, vol. 3657*, San Jose, CA, USA.

[17] Richardson, I. E. G. (2002). '*Transmission of Coded Video*', in: *Video Codec Design*, chap. *11*, J. Wiley & Sons.

[18] Chan, P. W., Lyu, M. R. & Chin, R. T. (2005). "A Novel Scheme for Hybrid Digital Video Watermarking: Approach, Evaluation and Experimentation", *IEEE Trans. on Circuits and Systems for Video Techn., vol. 15*, no. 12, 1638-1649, Dec.

[19] Ko, C. S., Kim, K. Y., Hwang, R. W., Kim, Y. S. & Rhee, S. B. (2005). "Robust audio watermarking in wavelet domain using pseudorandom sequences", 4th ACIS Int. Conf. *on Computer and Information Science*, 397-401, July.

[20] Wei, Z. H., Qin, P. & Fu, Y. Q. (1998). "Perceptual digital watermark of images using wavelet transform", *IEEE Trans. on Con-sumer Electronics, vol. 44*, no. 4, 1267-1272, Nov.

[21] Langelaar, G. C., Setyawan, I. & Lagendijk, R. L. (2000). "Water-marking digital image and video data. A state-of-the-art overview", IEEE Sign. Proc. Mag., *vol. 17*, no. 5, 20-46, Sept.

[22] Podilchuk, C. I. & Delp, E. J. (2001). "Digital Watermarking: Algo-rithms and Applications", *IEEE Sign. Proc. Mag., vol. 18*, no. 4, 33-46, July 2001.

[23] Anderson, R. J. & Petitcolas, F. A. P. (1998). "On the limits of steganography," IEEE J. Select. *Areas on Commun., vol. 16*, 474-481, May.

[24] Langelaar, G. C., Lagendijk, R. L. & Biemond, J. (1998). "Real-time labeling of MPEG-2 compressed video," J. *Visual Commun. Image Representation, vol. 9*, no. 4, 256-270, Dec.

[25] Wolfgang, R. B. & Delp, E. J. (1999). "Fragile watermarking using the VW2D watermark", Proc. *Electronic Imaging, '99, vol.* 3657, 204-213, Jan.

[26] Benedetto, F., Giunta, G. & Neri, A. (2005). "QoS provisioning and control in real-time interactive multimedia communication using digital watermarking", *Proc. of Tyrrhenian Interna-tional Workshop on Digital Communications, TIWDC*, Sorrento, Italy, July 2005.

[27] Wolf, S. & Pinson, M. (2002). "Video Quality Measurement Techniques", *NTIA Report*, 02-392, June.

[28] Watson, A. B. & Malo, J. (2002). "Video Quality Measures based on the Standard Spatial Observer", Proc. of Int. Conf. *on Image Processing, vol. 3*, 41-44, June.

[29] Xu, Y. & Zhou, Y. (2004). "H.264 Video Communication Based Refined Error Concealment Schemes", *IEEE Trans. on Consumer Electronics, vol. 50*, no. 4, 1135-1141, Nov.

[30] Sikora, T. (1997). "MPEG digital video-coding standards", *IEEE Signal Proc. Mag., vol. 14*, no. 5, 82-100, Sep.

[31] Rabiner, L. R. (1994). "Applications of Voice Processing to Telecommunications", Proc. *of the IEEE, vol. 82*, no. 2, 199-228, Feb.

[32] Chai, D. & Ngan, K. N. (1996). "Automatic Face Location for Videophone Images", Proc. of IEEE TENCON '96 - *Digital Signal Processing Appl., vol. 1*, 137-140, Nov.

[33] Ngan, K. N. & Rudianto, R. L. (1996). "Automatic face location detection and tracking for model-based video coding", 3rd Int. Conf. *on Signal Proc., vol. 2*, 1098-1101, Oct.

[34] Koenen, R. (1999). "MPEG-4 multimedia for our time", *IEEE Spectr., vol. 36*, 26-33, Feb.

[35] Challapali, K., Brodsky, T., Lin, Y. T., Yan, Y. & Chen, R. (2004). Yi, "Real-Time Object Segmentation and Coding for Selective-Quality Video Communications", *IEEE Trans. on Circuits and Systems for Video Techn., vol. 14*, no. 6, 813-824, June.

[36] Gao, Xio, Wu Gang, Miki, T. (2004). "End-to-end QoS provisioning in mobile heterogeneous networks", *IEEE Wireless Comm., vol. 11*, no. 3, 24-34, June.

[37] Alam, M., Prasad, R. & Farserotu, J. R. (2001). "Quality of Service among IP-Based Heterogeneous Networks", *IEEE Personal Comm., vol. 8*, no. 6, 18-24, Dec.

[38] Abedi, S. (2004). "Improved stability of QoS provisioning for 3G systems and beyond: optimum and automatic strategy selection for packet schedulers", *IEEE Int. Conf. on Comm., vol. 4*, 1979-1985, June.

[39] Islam, M. M., Murshed, M. & Dooley, L. S. (2004). "Enhanced cell visiting probability for QoS provisioning in mobile multimedia communications", Int. Conf. on Inform. Techn.: *Coding and Computing, vol. 2*, 258-262, 5-7 Apr.

[40] Eveno, N., Caplier, A. & Coulon, P. Y. (2001). "New colour transformation for lips segmentation", *IEEE 4th Workshop on Multimedia Signal Proc.*, 3-8, Oct. 2001.

[41] Benedetto, F., Giunta, G. & Neri, A. (2005). "A new color space domain for digital watermarking in multimedia applications", *IEEE Int. Conf. on Image Proc.*, ICIP Genoa (Italy), Sept. 2005.

[42] Gomez, G. (2002). "On Selecting Colour Components for Skin Detection", Proc. of 16th Int. Conf. *on Pattern Recognition, vol. 2*, 961-964, Aug.

[43] Tkalčič, M. & Tasič, J. F. (2003). "Colour spaces – perceptual, historical and applicational background", EUROCON 2003. *Computer as a Tool, vol. 1*, 304-308, Sept.

[44] Karungaru, S., Fukumi, M. & Akamatsu, N. (2001). "Detection of Human Faces in Visual Scenes", *Intelligent Information Systems Conf.*, the 7th Australian and New Zealand 2001, 165-170, Nov.

[45] Gilani, S. A. M., Kostopoulos, I. & Skodras, A. N. (2002). "Colour Image-Adaptive Watermarking", 14th Int. Conf. *on Digital Signal Proc., vol. 2*, 721-724 , July.

[46] Chai, D. & Ngan, K. N. (1999). "Face Segmentation Using Skin-Colour Map in Videophone Applications", IEEE Trans. *on Circuits and Systems for Video Techn., vol. 9*, no. 4, 551-564, June.

[47] Fritsch, J., Lang, S., Kleinehagenbrock, M., Fink, G. A. & Sagerer, G. (2002). "Improving Adaptive Skin Colour Segmentation by Incorporating Results from Face Detection", *IEEE Int. Workshop on Robot and Human Interactive Comm.*, Berlin, Germany, Sept.

[48] http://www.comlab.uniroma3.it/ens/video.zip zipped WEB folder.

[49] Kim, D. H., Do, H. C. & Chien, S. I. (2005). "Preferred Skin Colour Reproduction Based on Adaptive Affine Transform", *IEEE Trans. on Consumer Electronics, vol. 51*, no. 1, 191-197, Feb.

[50] Lee, E. J. & Ha, Y. H. (1997). "Automatic Flesh Tone Reappearance for Colour Enhancement in TV," IEEE Trans. *on Consumer Electronics, vol. 43*, no. 4, 1153-1159, Nov.

[51] Mendes, P., Schulzrinne, H. & Monteiro, E. (2002). "Session-aware popularity-based resource allocation for assured differentiated services", *IEEE Commun. Mag., vol. 40*, no. 9, 104-111, Sep.

[52] Qingwen, L., Xin, W. & Giannakis, G. B. (2006). "A cross-layer scheduling algorithm with QoS support in wireless networks", *IEEE Trans. on Vehic. Techn., vol. 55*, no. 3, 839-847, May.

[53] Wenjun, Z., Junqiang, L. & Xinhua, Z. (2006). "Security for multimedia adaptation: architectures and solutions", *IEEE Multimedia, vol. 13*, no., 68-76, April-June.

[54] Ahmed, T., Mehaoua, A., Boutaba, R. & Iraqi, Y. (2005). "Adaptive packet video streaming over IP networks: a cross-layer approach", IEEE J. *on Selected Areas in Commun., vol. 23*, no. 2, 385-401, Feb.

[55] Matsumoto, H. & Mizui, K. (2000). "A Prototype of 2-Way Spread Spectrum Communication System for Inter-Vehicle Communication", IEEE Canadian Conf. *on Electrical and Computer Eng., vol. 2*, 1047-1051 March.

In: Videoconferencing: Technology, Impact and Applications ISBN: 978-1-61668-285-9
Editor: Adam C. Rayler, pp. 35-55

Chapter 2

VIDEO-MEDIATED COMMUNICATION AND IMPRESSION FORMATION: AN INTEGRATIVE REVIEW

Chris Fullwood and Mandy Finn
Psychology Department, University of Wolverhampton, Wolverhampton, U.K.

INTRODUCTION

Although the technology to support site-to-site videoconferencing has been practicable for over 70 years (Fullwood, 2007), video-mediated communication (VMC) has failed to shape the social world to the same extent as the telephone. Up until recently, the corporate sector has been the dominant consumer of VMC. It is only in recent years that video-mediated communication has penetrated the personal sector on a global scale. For over a decade now, Internet Protocol based videoconferencing has provided the general public with a reasonably priced alternative to more conventional means of distanced communications. VMC is an emerging market and is likely to continue to grow as technology advances. In a survey of 1,450 respondents, Rainie (2005) found that 16% of American adult Internet users had used a web cam to view another person or place. This translates into approximately 21 million Americans who had utilised the Internet for VMC purposes in 2005. Currently, there are estimated to be around 251 million Americans accessing the Internet, approximately 74.4% of the total population (Internet World stats, 2009), so we would expect this figure to have increased significantly. Rainie suggests that the 2005 figures represent an increase in web cam usage from previous years and this can be accredited to two specific factors: web cams are now cheaper and easier to set up, and broadband speed makes it easier to display images. As video-mediated technologies become a more pervasive aspect of our daily lives, questions relating to the psychological impact of using the technology become ever more prevalent.

There are many advantages that VMC should offer over the telephone. For example, VMC presents the opportunity of duplicating many of the benefits associated with interacting in a co-present environment (Sellen, 1997). According to Campbell (1998), videoconferencing systems closely imitate face-to-face meetings, without encumbering the user with the financial and physical costs of travelling. Unlike audio-only communications,

VMC permits participants to perceive the reactions of their colleagues. According to the 'non-verbal hypothesis' (Whittaker, 1995; 1996), access to non-verbal cues should allow users to respond with greater comprehension and sensitivity. For example, one should be able to notice if another person is smiling or frowning (Campbell, 1998), and this may have a profound impact on the exchange. Furthermore, Isaacs and Tang (1993) indicate that adding a visual channel to remote communication produces many benefits for users. These include: allowing listeners to signal their understanding (or lack of) in the non-verbal channel; speakers can supplement verbal information with gestures; listeners can respond to speakers in an entirely visual capacity (for example, head-nodding); attitudes can be communicated with posture and facial expressions; and the significance of conversational pauses can be interpreted more easily. For all of these reasons, one would expect video-mediated communication to be a more satisfactory, effective and efficient mode of interaction than audio-only communication.

Although VMC is undoubtedly a more cost-effective alternative than travelling great distances to a meeting (Strom, 1997), there are a number of second-order effects associated with the use of the technology, which are yet to be fully understood (Ferran-Urdaneta, 2001). The fact is, although VMC is a close approximation of face-to-face communication, it is certainly not the same as meeting someone in person. Hence, there is a problem when users of the technology treat face-to-face communication as the benchmark and assume that video-mediated communication is equivalent to co-present interaction. To date, much of the literature has focused on the manner in which humans can adjust to and apply emerging technologies in their everyday lives. In particular, research has adopted a task-based approach, for example identifying the manner in which users perceive the effectiveness of VMC for problem-solving and negotiation style tasks (for example, see Anderson, 2006). However, there has been far less consideration of the psychological impact of using VMC. Although VMC is utilised for a number of different purposes and many of these are often very specialised (e.g. telemedical applications), there are many applications for which it is necessary to get a feel for another person (for instance, interviewing from a distance). It would seem crucial then that the technology does not impede our ability to form impressions of others. This may be particularly important if initial contact is made via VMC, or if comparisons are made between those met via video and those met face-to-face. This review will therefore consider empirical research on VMC and impression formation in a variety of contexts. It will also attempt to explain any differences between the manner in which people make inferences about others in video-mediated and face-to-face settings. Finally, recommendations for improving video-mediated interactions will be offered. Before this, however, the review will briefly reflect on several of the major barriers to successful video-mediated communication that have been recognised in the literature. This is important, as an understanding of the technological constraints of VMC will help illuminate any differences that have been observed in the impression formation process when comparing VMC and face-to-face communication.

Barriers to Successful VMC

Technological Barriers

There are considered to be a number of barriers to successful VMC and the majority of these can fit into either a 'technological' or 'non-verbal' category, whereby 'non-verbal' refers to the user's ability to adequately express and accurately read non-verbal signals when using video-mediated technologies. In the 'technological' domain, one major barrier is the quality of the audio signal. Many researchers concur that good audio quality is necessary for videoconferencing to thrive (for example, Angiolillo et al, 1997). However, this will be dependent upon good room acoustics and microphone placement, which may not always be achievable (Angiolillo et al, 1997). Difficulties in hearing other members of the exchange can result in unnecessary repetitions and misinterpretations, which inevitably hinder communicative performance. The importance of a high quality audio signal has been demonstrated in a number of studies. For example, Watson and Sasse (1996) noted that participants judged a video signal as being better when the quality of the audio signal increased. It seems likely that face-to-face communication is the yardstick for many users, drawing inevitable, and often unrealistic, comparisons.

One of the most significant barriers to successful VMC is audio delay, in other words, when signals (audio and visual) from one site do not reach the other location straight away. Visual and audio signals may also be processed at different speeds, resulting in a lack of synchronisation. Overlooking lip synchronisation can be harmful to communication (Angiolillo et al, 1997), however audio delay is considered to be much more frustrating to users (Tangs and Isaacs, 1992). The length of delay varies from one system to another, dependent on the type of algorithm used, and within a system, dependent on the amount of information that needs to be compressed at any one time (Angiolillo et al, 1997). When delay occurs, people are often unaware of it and experience it generally as confusion. In many cases problems with communication are attributed to the other member of the exchange and not to the technology itself (Kitawaki, Kurita & Itoh, 1991). In other words, users will often perceive the other member as being slow to respond and will fail to recognise that it is the video system that is actually causing the hold-up. O'Malley et al (1996) compared task performance on the 'Map Task' (Brown et al, 1984) with video-mediated conditions with delay and video-mediated conditions without delay. Delay of approximately 500 milliseconds was accomplished using "off–the-shelf videophones" with an analog signal. When completing the task with a delay in video, over 50% of a speaker's turns were interrupted and task performance was poorer. Human communication is reliant on refined timing and therefore any disruption to this process is likely to interfere with the flow of communication. Delay in video has an adverse effect on communication and task performance and therefore should be avoided in the design of remote communication systems (O'Malley et al, 1996).

The size of the monitor/screen is also an important technological consideration. Videoconferencing is used principally for the observation of speakers, but it is evident that as the number of people on the screen increases, each individual becomes more difficult to see. According to Tiffin and Rajasingham (1995) videoconferencing systems are optimally employed with the use of a large screen to show the person who is talking close-up. This therefore makes the most of any non-verbal signals that need to be transmitted. Small screens are valuable in that they are portable, and because lower resolution images look clearer (Angiolillo et al, 1997). Users however prefer large screens and this can be ascribed to a desire for social co-presence. A larger screen provides an intuitive feeling that the distanced participants are physically present in the same location (Angiolillo et al, 1997). This theory is borne out by De Greef and Ijsselsteijn (2001), who indicate that life-sized, broadband video results in improved perceptions of social co-presence. Even though there appears to be some advantage to using larger screens, video display sophistication however must have limits due to bandwidth, cost and practicality. Therefore, a trade off is required for the sake of convenience, but also to safeguard the advantages of using VMC in the first place (Angiolillo et al, 1997).

When it comes to the optimal set-up, Martino et al (2006) recognise that factors such as price, convenience and the resources available will unavoidably impress limitations on what users can accomplish. However, they also suggest that, where possible, users should endeavour to use a system with high-quality audio and video, minimal delay and a clear view of all speakers taking part in the exchange. The purpose for which the system is being used for is also an important consideration. Achieving an optimum set-up would be less valuable for communications that are recreational in nature. Martino et al (2006) also suggest that in making the most of the technology, users should be aware of the restrictions in order that they may adapt effectively.

Non-Verbal Barriers

Although it is difficult to consider VMC as a homogenous entity, due to the many different possible set-ups and differences in quality between systems, a number of researchers have noted that there is an increased difficulty associated with reading and expressing non-verbal information comparative to face-to-face communication. Moreover, in many cases this seems to be an issue irrespective of the set-up or quality of the system being used. The major concern would seem to be the inability of users to express the normal modulations of eye gazing behaviour. Although users undoubtedly have access to visual images, it would seem that natural eye contact is compromised in VMC. The eye contact problem is due to camera placement, specifically that it is mounted above the screen. It is possible to remedy this predicament by altering the position of the eyes of the viewed partner to a particular area of the monitor and then arranging for the position of the camera to align with this area using mirrors as in the Xerox EuroPARC Video Tunnel (Gaver et al, 1992). Research by Doherty-Sneddon et al (1997), however indicates that participants overuse the visual channel when Video Tunnels are used in communication. In essence, participants spent far longer looking their conversational partner in the eyes comparative to a face-to-face equivalent context. This over-gazing effect may in turn hamper our ability to process cognitive information and plan speech effectively (Beattie, 1981), particularly as over-gazing can lead to increased arousal, which has been associated with poorer cognitive performance (see Yerkes-Dodson, 1908). Therefore, the benefits of being able to make eye contact via VMC may be offset by the use of considerably more speech to achieve an analogous level of task success. The over-gazing effect observed by Doherty-Sneddon and colleagues may also be explained with reference to equilibrium theory. According to Argyle and Dean (1965) there are a number of cues that we use to express intimacy (for example proximity, touch and eye contact). In situations where some of these cues are restricted, individuals may compensate by expressing other available cues more freely. For example, we are said to make more eye contact with others when they are further away from us. In this sense, participants may have compensated for a lack of proximity by over-gazing. Technological constraints in VMC place limits on our ability to make eye contact and use gaze awareness. These limitations are contributing factors towards the failure of VMC to fully depict non-verbal signals and according to Angiolillo (1997) is one of the key factors that prevents video from providing value to remote conferencing.

The same level of non-verbal interaction found in face-to-face communication does not occur in the use of video-mediated technologies and has a number of effects. As well as the problem associated with gazing behaviour, it is evident that many systems use only headshots to take advantage of subtle non-verbal signals communicated by the face. However, in using headshots many of the benefit that non-facial non-verbal behaviours such as posture, and gesture have to offer are eliminated (Videotalk, 1998). Indeed, O'Malley et al (1996) provide some evidence that video-mediated technologies are more effective when access to the head and shoulders is provided as opposed to a headshot only. As far as facial expressions are concerned the major foreseeable problems concern delay and the size of the image. In instances where the size of the image is small, this makes it increasingly difficult to pick up important non-verbal signals. Both 'non-verbal' and 'technological' barriers are likely to impact upon the communication process. Although 'perfect' videoconferencing may be achievable in terms of good quality images and no delay, this may only be in a limited number of specialised systems, which are not necessarily available to the general public. The majority of users are therefore likely to be working with imperfect technology. Furthermore, the user's ability to express and interpret non-verbal behaviours is expected to be an issue, irrespective of the quality of the system. In particular, the eye contact problem has yet to be resolved satisfactorily. Although these may not be significant concerns for those who use the technology for more recreational pursuits, problems may occurs when users need to work collaboratively to make important decisions. The review will now consider research on impression formation in a variety of video-mediated communication contexts.

VMC and Impression Formation

The manner in which we gather impressions of other individuals has been of special interest to researchers for many years now. When making inferences about another person, the perceiver is likely to draw conclusions that are based on the sender's use of non-verbal and verbal social cues. Furthermore, it is argued that non-verbal cues (for instance aspects of appearance and the use of eye contact) are particularly salient when it comes to making judgements about others (Argyle, 1994). Research suggests that humans can make accurate impressions of others based on only a modest amount of information (Bar et al., 2006). We often form impressions rapidly and first impressions are fairly persistent (Asch, 1946). As VMC is often utilised in situations where it is important to gather impressions of another individual (e.g. in the courtroom), it would seem crucial that accurate and favourable impressions are formed. Research by Fullwood (2007) illustrates that impressions of others can be more negative when communication takes place via video. Participants were asked to judge co-participants on a number of personality variables using a novel but simple mind-reading task. This task was used in particular because it distracted participants away from the real aim of the research. Half of the participants completed the task face-to-face, whereas the remaining participants completed the task via VMC. Participants were rated less favourably by their partners in terms of likeability and intelligence when the task was completed via video. These findings are concerning, particularly as users are unlikely to be aware of the manner in which the technology influences their decisions. VMC is increasingly being utilised in business and courtroom settings. In each of these cases gathering a sense of another person is imperative. The consequences of drawing less accurate and more negative impressions may be far-reaching, not only for the individual who is being assessed, but also for the overall perceptions of the usefulness of the technology. The review will now consider research on impression formation in these two specific VMC contexts.

VMC and Impression Formation: Business Applications

With an ever-expanding global economy, organisations are increasingly utilising video mediated technologies for business purposes. Essentially, technology is changing the way that we work. In a business environment VMC may be used between companies for communication purposes, but also within companies, for example in learning, mentoring or coaching contexts. Colky and Young (2006) argue for the benefits of virtual mentoring insofar as it aids in the reduction of company costs, provides increased access for workers and does not place limitations on space, time or location. Making generalisations about the influence of mediated communication is tricky as media fluctuate in terms of sequentiality, visibility, co-presence and simultaneity (Clark and Brennan, 1991). However, in face-to-face context communication is always likely to be more fluent as direct interaction produces a logical sequence in discussions (Hambley, O'Neil & Kline, 2007; Silvester et al., 2000). In a video-mediated communication context, delay may impact upon the timing of utterances, meaning that communication is more contrived. However, irrespective of the technological limitations of the system under use, there are many persistent differences between VMC and face-to-face communication. For example, in all forms of VMC, the users are communicating from a distance and are therefore not in a co-present environment, and true eye contact is unlikely to be achieved.

One business application that has received popular attention is the use of VMC for interviewing. Interviews at a distance are common practice in the business world and in many cases companies use this method to screen candidates. Although interviewees are unlikely to be at a disadvantage if they are being directly compared with others being interviewed by the same method, a problem may arise when comparisons are made with face-to-face interviewees. Chapman and Webster (2001) found that face-to-face candidates were rated more favourably in terms of their use of language and non-verbal communication than candidates interviewed by video. Chapman and Rowe (2002) suggest that applicants interviewed by video feel disadvantaged and in turn they have less positive attitudes regarding companies who employ this method in their selection process. Moreover, Chapman et al (2003) suggested that face-to-face interviews were perceived as fairer than video-mediated interviews. Furthermore, applicants were more likely to receive an offer after a face-to-face interview. Straus et al (2001) however found that face-to-face and video-mediated interviewees were rated equally as well. Interestingly, telephone interviewees were rated more favourably than face-to-face interviewees, and this effect was stronger for applicants rated as less physically attractive. However, the interviewees had less positive attitudes about using video-mediated technology for interviewing and the interviewers were perceived to be less friendly in the VMC condition compared to the face-to-face condition. Overall these studies suggest that video-mediated interviewees are likely to be at a disadvantage and this may be because they are perceived less favourably than their face-to-face counterparts.

A study by McCulloch and Oborne (1999) tested whether increased bandwidth and resolution would impact upon perceptions of interviewees. There were eight conditions in total, with four bandwidth conditions (56k, 128k, 224k and 384k), each with two levels of resolution (high resolution = 352x288 pixels; low resolution = 176x144 pixels). Their findings indicate that even though as bandwidth increased so did perceptions of telepresence (i.e. the subjective feeling of being in the same location), perceptions of the interviewee did not differ. An overall impression formation rating was gathered by taking factors such as honesty, warmth and confidence into consideration. In general terms, impression formation ratings were modest for all bandwidth levels (approximately 3 on a 5 point likert scale). On the one hand this is good news for interviewees using video, as the quality of the system is unlikely to have a negative impact upon impressions. This means that those who are interviewed with a low quality system are not going to be disadvantaged comparatively to those who are interviewed with a high quality system. However, the results also suggest that, irrespective of the quality of the system, interviewees are likely to be perceived less favourably generally. This means that even with advances in technology, face-to-face interviewing is likely to give you the best chance of being selected for a job. Nonetheless, it may be worth considering the possibility that some individuals would be advantaged by video-mediated communication in this context. Those who are shy or nervous, for example, may find it easier to project themselves via video because of a reduction in social pressures associated with face-to-face contact. This line of enquiry is however untested and would be worthy of further investigation.

VMC technologies may also be used by businesses for remote collaboration purposes. In these instances, reaching mutual consensus is paramount; therefore it is important that interlocutors can establish mutual trust. In this sense, it may be important for co-communicators to work out whether or not another individual is being truthful. Kiesler and Cummings (2002) suggest that collaborating at a distance is more difficult that collaborating face-to-face. This may be because 'trust needs touch' (Handy, 1995). Indeed, Handy (1995) argues that as an organisation becomes more 'virtual' there is a greater need for face-to-face contact because only personal contact can make a shared commitment real. Evidence to support this notion comes from Horn (2001), who found that participants were poorer at detecting deception and picking up on subtle non-verbal cues when communication took place via a video link in a collaborative context. Empirical evidence also suggests that geographically distributed work-teams are more likely to experience high levels of conflict. Hinds and Bailey (2003) propose that such conflict is a direct consequence of having to rely on the technology for communication as well being physically distant from co-workers. It may also be the case that physical distance and reduced social cues mean that individuals feel less inhibited and therefore more able to disagree with others. Indeed, Suler (2003) discusses the 'online disinhibition' effect in computer-mediated communication, and a similar effect may occur in VMC. Rockman & Northcroft (2008) report on the impact of using leaner media on defection, deception and trust. In their study, a social dilemma exercise was utilised, where it was necessary for participants to cooperate in order to complete the task successfully. Participants completed the task in either a face-to-face, video-mediated or computer-mediated condition. Participants felt less confident in trusting their co-participants and experienced an increased difficultly in identifying others' motivations and intentions when communication took place via leaner media. Furthermore, because participants felt less able to trust their partners, they were more likely to be deceptive and less likely to cooperate. Overall, there appears to be compelling evidence that, for specific business applications, the use of VMC may have a negative influence on person perceptions. The use of video-mediated technologies would seem to interfere with our ability to gather a sense of who someone is, as well as resulting in more negative impressions of others generally. Interviewees are likely to be perceived less favourably if they are interviewed via video in comparison to candidates interviewed face-to-face. Collaborative teams are also liable to find it more difficult to establish trust in a VMC context and may also engage in more deceptive and defective behaviours.

VMC and Impression Formation: Courtroom Applications

Video-mediated technologies are increasingly applied in courtroom contexts. The benefits of this strategy are relatively obvious; emotional and social pressures are weakened and therefore presenting evidence should be a far less daunting experience (Doherty-Sneddon and McAuley, 2000). Furthermore, many individuals would not present evidence if the only option were to do so in the courtroom. Although evidence suggests that vulnerable witnesses find it easier to present evidence via video (for instance, Doherty-Sneddon and McAuley, 2000), there is also corroboration for the notion that jury perceptions of a witness are more negative via VMC. Landström and colleagues (2005) present persuasive evidence in a series of studies to suggest that video link witnesses are perceived less favourably in terms of honesty and appearance comparative to face-to-face witnesses (Landström et al, 2005; Landström et al, 2007). When considering the cases of video link child witnesses, Davies and Noon (1991) found that judges were less likely to be influenced by emotional considerations, and were therefore less empathetic towards the witnesses. Evidence from Fullwood et al (2008) indicates that participant jury members found it more difficult to emotionally engage with an eyewitness and considered their statement to be less credible via video comparative to face-to-face testimony. However, the mode of communication did not impact upon the jury's decision to convict the accused on the basis of the evidence provided. The witness did however present testimony relating to a fairly innocuous incident, and one may only speculate as to the potential findings if the testimony were related to a more severe crime.

Additional evidence from the Scottish Office (1995) implies that lawyers are particularly critical of the video-mediated technologies being employed in the courtroom, and this can be in part credited to the increased difficult associated with being able to build a rapport with the witness. The report also concluded that the jury found it more difficult to gather an accurate impression of the witness and that the video served as a barrier to successful communication.

Although research findings seem to suggest that presenting evidence via video may have a negative impact on impressions, there are also findings in the literature that suggest that mediation does not influence perceptions of credibility (for example, Orcutt, 1995; and Goodman et al, 1998). Furthermore, Taylor and Joudo (2005) found that perceptions of guiltiness were unaffected by mode of testimony, even when the testimony related to a severe crime.

Evaluating the Difference between VMC and Face-to-Face Impression Formation

Evidence suggests that, in very specific contexts (e.g. in the courtroom and for interviewing) as well more generic contexts, VMC can result in more negative perceptions of users on a number of personality variables. For instance, comparative to face-to-face participants, VMC participants have been reported to be less likeable, less intelligent, poorer users of language, less trustworthy, less honest and less emotionally engaging. There are a number of important differences that distinguish face-to-face interaction from VMC and these are likely to affect the impression formation process. These distinctions will be discussed further below.

Distorted, Missing or Attenuated Visual Signals

Communication media differ in terms of their richness (i.e. how many cues are available to the communicators). Verbal and non-verbal cues are more easily communicated with the use of richer media, and for this reason they promote more 'natural' language (Daft and Lengel, 1986). As many VMC systems are imperfect, resulting in delay or poor visual and audio quality, the quality and accuracy of any message transmitted may be degraded. Consequently, the poorly communicated message may be attributed to the sender and not to the technology, resulting in more negative views (Rice, 1993). This may be particularly important in an interview context, specifically as interviewees are likely to evaluate their performance on the basis of the feedback given to them during the interview (Martin & Nagao, 1989). Accordingly, when communication barriers interfere with the feedback process, interviewees may find it more difficult to judge their performance and this may impact upon their confidence (Chapman et al, 2003). The type of set-up may also play an important role. In VMC, cameras are commonly placed above the monitor and often in high positions. Huang (2005) manipulated the angle and distance of the camera in relation to the user. Participants interacted in pairs and completed a decision-making task on Arctic survival. Although performance on the task was unaffected by camera setting, participants who appeared 'taller' as a result of the camera setting, made more arguments and decisions and were perceived to be more influential in the task than those who appeared 'smaller'. Huang (2005) presents striking evidence that the use of a monitor and camera as a proxy for co-present interaction distorts the visual image and in turn this influences perceptions of others.

Heath and Luff (1991) and Rutter (1987) argue that non-verbal cues in VMC fail to impact upon receivers in the same way as they do in a face-to-face context. According to

Heath et al (1995) "the resources upon which a speaker ordinarily relies to shape the ways in which a co-participant listens and attends to the talk appear to be interfered with by the technology" (Heath et al, 1995, pg. 174). The monitor distorts the shape and movement and destroys "the relative weighting of different aspects of an individual's conduct" (Heath et al, 1995, pg. 174). In essence, the non-verbal bodily activity produced by an individual looks very different on the monitor than it would if it were perceived in a co-present environment and this is likely to influence the manner in which they are perceived. Non-verbal cues are weighted heavily in the impression formation process (Argyle, 1994). In circumstances where they are distorted, attenuated or removed from communication, people are more likely to be assessed on the facts and less likely to be judged on their personalities (Williams, 1977; Morley and Stephenson, 1969).

As well as being distorted, in many VMC contexts non-verbal cues may also be unavailable. For example, when users present a head and shoulders shot of themselves, important non-verbal cues that are communicated by the arms and hands will not be visible. Dennis and Kinney (1998) suggest that there is a general consensus that leaner media (i.e. with fewer available cues) are more task-focused, more impersonal and less friendly. Non-verbal cues play an important role in message production, assessment and comprehension. They also aid in the regulation of turns at talk and intimacy expression (Derkes, Bos & Grumbow, 2007; Driskell & Radtke, 2003; Hidalgo-Barne & Massaro, 2007). There is also an abundance of research, which demonstrates the influence of visual cues in the communication of interpersonal attitudes or effect. In support of this notion, Grahe and Bernieri (1999) provide evidence for the significance of non-verbal cues in judging rapport. On the basis of transcripts and audio and video records of interactions, participants were asked to judge the rapport of interacting dyads. When non-verbal, visual information was available, participants more accurately judged the rapport of the conversational participants. Furthermore, Bernieri & Rosenthal (1991) suggest that in face-to-face interactions, rapport is established implicitly through the synchronisation of emotional expression and gestures. Establishing mutual understanding in a face-to-face context occurs within a tight frame (Condon, 1982) and this is likely to be below the frame rate for the majority of VMC set-ups (Parkinson, 2008). Therefore, building a rapport using technology is likely to be very different than building a rapport face-to-face. Driskell and Radtke (2003) illustrate that when gestures are removed from the communication process, speech production and comprehension were negatively affected. This is because gestures help speakers to produce statements and listeners to understand them. Driskell, Radtke & Sales, (2003) argue that participants are less able to judge if understanding has occurred when non-verbal cues are removed from communication. Head-nods, for example, are effective in allowing speakers to recognise that their messages are being understood.

Interpretation of Emotional Information

When it comes to gathering impressions of others, an individual's emotional state would be a determining factor in how we react to and feel about them. On an interpersonal level, there is an ongoing process of reciprocation in which our emotions affect and are influenced by the emotional expression of others (Parkinson, Fischer, & Manstead, 2005). If an individual presents a negative emotional display for instance, we are likely to have a less favourable attitude towards them, which in turn may lead to a less favourable impression of us. Furthermore, Asch (1946) describes how the traits of coldness and warmth are weighted heavily and that once we ascribe someone to one of these categories, we make mental shortcuts to other aspects of their personality. For example, someone who is perceived as cold may also be seen as unapproachable or unfriendly. During real-time, face-to-face communication, non-verbal cues associated with emotional expression can be observed directly by the receiver. However, in video-mediated settings, it is not always possible to perceive non-verbal cues as they may be missing, attentuated or delayed (Parkinson, 2008) and this is likely to influence the interpretation of emotions. To highlight this notion, we should consider a situation in which an individuals expresses anger. If the sender of a message expects an angry retort to his/her message but the reply is delayed, the level of anger expressed in reply may appear somewhat disproportionate (Parkinson, 2008). Interactions are sequential and co-ordinated, for example a question is followed by an answer, and this has been referred to as 'relation alignment' (Parkinson, Fischer & Manstead, 2005). When the technology interferes with this process, interpretations of messages are likely to be very different, and therefore this may influence how we feel about the person sending the message. Further support for the notion of the manner in which video-mediated communication influences the perception of emotional messages comes from Parkinson & Lea (in press). In their study, participant pairs took part in communications about celebrities via VMC. The pairs discussed celebrities that they liked and disliked under two conditions: a disagreement condition (one of the participants liked the celebrity and the other did not) and an agreement condition (both participants disliked the celebrity). Furthermore, one group experienced minimal transmission delay and the other group a small transmission delay. The results suggest that participants found it more difficult to establish a rapport when there was a delay and this is a likely consequence of a lack of immediate feedback. It is evident that the different forms of communication media have an impact on how emotion is conveyed or interpreted and this would likely influence the impression formation process. If some aspects of emotional expression seem more intense in the video medium, it is easy to see how this may lead to more negative reactions, particularly if those emotions are negative to begin with. Furthermore, a difficulty in establishing an emotional rapport with another may lead to more negative impressions of that person.

The Eye Contact Problem

As previously discussed, one of the major issues associated with VMC is the difficulty in replicating eye contact. Eye contact cannot be truly preserved in VMC because the individual cannot look in to the camera (to give the impression of eye contact) and at the image of the other person on the monitor at the same time. The most common compromise is to place the camera above the screen. This results in the perception of individuals casting their eyes in a downward direction, however this is more favourable to other positions, as gazing upwards makes people appear suppliant and gazing sideways makes people appear deceitful (Kenyon et al, 1985). There have been several novel solutions to preserving eye contact including video tunnels and the clearboard project (Ishii & Kobayashi, 1992), which superimposes the video of the other participant on a common whiteboard while a camera preserves eye contact using half-silvered mirrors (Angiolillo, 1997). These particular systems, however, are not readily available to the average user, and therefore this problem with eye contact is still a persistent one for most users of VMC. Storck and Sproull (1995) argue that less favourable impressions in VMC are a likely consequence of a lack of direct eye contact. Indeed, they suggest that VMC users are often perceived as being unfriendly for this very reason.

Gazing behaviour is central to the communication of attitudes and feelings. Speakers are said to gaze more frequently when talking, and this is thought to be a direct attempt at persuasiveness (Mehrabrian & Williams, 1969). Speakers may also look others in the eye more regularly when being deceptive (Burns & Kintz, 1976), dominant (Exline, 1971; Ellsworth, 1975) or sycophantic (Pelligrini, Hicks & Gordon, 1970). As gaze is used to communicate feelings and attitudes, it is not surprising that people are judged on their use of gaze. Indeed, individuals are judged as evasive when they look infrequently at others and friendly and sincere when they look at others more often (Kleck and Nuessle, 1968). Those who directly gaze at others are judged to be more intelligent (Tankard, 1970; Larsen and Shackelford, 1996). Those who avert gaze in communication are also evaluated as being less credible and more likely to be hiding something (Hemsley and Doob, 1978; Kraut and Poe, 1980). It could be argued therefore that an inability to communicate the normal modulations of eye gazing behaviour is liable to influence our impressions of others (Fullwood, 2007). We might find it more difficult to detect whether someone is being truthful for example as eye contact is a particularly salient cue in this respect, and this would be of particular importance in a courtroom setting. The perception that someone is looking away constantly (which would be common in VMC communications) may also give the impression that this person is disinterested, shy or evasive. However, it should be noted that just because gaze aversion is a cue for detecting dishonesty, does not necessarily imply that people interpret this cue accurately. Indeed, individuals may avert gaze for many other reasons than because they are hiding the truth, for example they may be shy or disinterested in what the speaker is saying. Irrespective of this, one might expect the perception of gaze aversion, which is loaded with negative associations, to impact unfavourably upon impressions of others.

Appearance Consciousness

Another important factor to consider when explaining why VMC impressions are often more negative than face-to-face impressions is appearance consciousness. It is likely, especially for inexperienced users, that communicating via video creates feelings of being observed (Fullwood, 2007). This may be particularly pertinent in the use of specific set-ups, for example the picture in a picture (PIP) set-up, in which users can not only see their co-participants, but also themselves. Evidence for the effects of appearance consciousness on impression formation comes from a series of studies by Wegge (2006). In the first study, the behaviours of students during a simulated oral examination were observed. Students were noticed to perform poorer when a large image of themselves was displayed. In the second study, participants took part in a consultation episode, which was considered to be less threatening. The results suggest that participants expressed anger more intensely when they could see their own picture. Furthermore, participants performed more poorly on the task with the counselling quality being compromised when a picture could be seen. Therefore, both studies imply that task performance can be hindered when one's own picture is displayed during a videoconference. Wegge (2006) suggests that this type of set-up deserves more attention in the academic literature. It would seem possible then that negative perceptions of users may also be caused by individuals feeling uncomfortable, nervous or monitored and therefore behaving differently on camera. Furthermore, this is also likely to be extended to set-ups in which the users cannot see themselves on screen.

RECOMMENDATIONS

An awareness of the limitations of VMC is essential, particularly in the courtroom, where the consequences of making the wrong decision can be catastrophic. Those who use VMC for impression formation purposes could therefore be advised to take into consideration the likely impact of the technology on person perceptions (Alexander et al, 1999). This could, for example, involve explaining the problem with eye contact. For instance, this may involve explaining that the appearance of someone looking away does not necessarily reflect a lack of interest, but is a product of the technological limitations of the system. However, this is an untested hypothesis, and there may be the possibility of users overcompensating when such advice is given. Users should also speak clearly, pause for reflection, allow time for delay (if appropriate) and indicate clearly when they have finished their turn (Alexander et al 1999). This advice would seem appropriate, as research implies that it is more difficult to manage interactions in video-mediated communication than in face-to-face communication (Doherty-Sneddon et al, 1997). Indeed, through indicating clearly when a turn is finished, users can compensate for the lack of non-verbal cues (for example eye gaze and head turning) that are used in the turn-taking process. It is likely that a more successful communicative exchange would lead to more favourable impression of the assessed individual generally. In the visual domain, users should take into account that subtle expressions will more than likely be lost. They should also avoid the use of excessive hand movements, however hand signals should be used to draw attention to who is speaking. O'Conaill et al (1993) also suggest that remote collaborators should attempt to choose appropriate communication technologies for the task at hand and ensure that certain types of tasks, for example negotiation, should always be resolved if possible in a face-to-face setting. Although it is clear that improving the quality of a video system (low lags, full duplex channels) leads to improvements in communication, as evidenced by the superiority of LIVE-NET over ISDN in O'Conaill et al's research (1993), the LIVE-NET results also suggest that improving these properties alone will not exactly reproduce face-to-face interaction.

There are also a number of strategies that can be employed to increase the possibility of gaining favourable impressions from others. One such strategy may be particularly practical for courtroom applications and involves the closing of intimacy through initial face-to-face contact. According to Abbott et al (1993), videoconferencing needs to be placed alongside physical contact in order for distanced meetings to be effective. It can also be argued that one of the factors that influences negative impressions is the lack of co-presence in video-mediated meetings. If we feel distanced from others, our impressions of them are likely to be different than when we feel proximate. This particular strategy has been used with some success previously. For example, Rocco (1998) found that participants were better at establishing trust with other group members if they had met beforehand in a face-to-face setting. Derrer et al (2006) noted that impressions of job suitability, honesty, friendliness and employability were better when an applicant had been met face-to-face briefly before a video-mediated interview. However, Fullwood et al (2008) found that this method did not alter jury perceptions of an eyewitness. Fullwood et al (2008) did however only use a very brief introduction to the witness, and therefore one might speculate as to the effect of longer contact in the face-to-face domain. Clearly, this method may not always be feasible, particularly in an interview context where VMC may be employed precisely for the reason that applicants are unable to travel to the location at where the interview is being held. However, what it does suggest is that, where possible, this method may be useful in bridging the intimacy barrier, resulting in improved perceptions of co-presence and therefore more receptive attitudes to others. However, whether these impressions are actually more accurate is another matter altogether and should be investigated further.

Another similar strategy that may be employed is the use of an icebreaker. This can be in the form of a team-building exercise for example and may help to reduce the anxiety associated with inexperience and appearance consciousness as well as focusing attention away from the novelty of using the technology for group interactions. This strategy has been recommend by Alexander et al (1999) and was tested empirically by Fullwood et al (2007). After taking part in a 'getting to know you' style task, collaborative pairs completed a map reading task more efficiently. For instance, participants completed the task quicker, in fewer turns and with fewer interruptions. It is argued that the initial icebreaker task helped pairs to establish common ground and mutual understanding, which lead to a more efficient performance on the task. Although somewhat speculative, one may expect improved mutual understanding to lead to a convergence of attitudes, which may be reflected in improved attitudes towards others.

Further strategies can be employed to make the best possible use of the technology in order to communicate more successfully. As a lack of eye contact in a VMC setting can be problematic and video tunnels are not freely available, users can be advised to look directly into the camera for emphasis. Although this does not replicate eye contact, it will give the impression to the other member of the exchange that he/she is being looked at directly in the eyes. This strategy was employed by Fullwood and Doherty-Sneddon (2006) and was shown to help participants to recall more information about a sales pitch. Although this is not impression formation per se, it is evidence that one of the functions of eye gaze can be replicated in VMC. Therefore, there is the possibility that the impression formation functions of eye contact demonstrated in a face-to-face context may also be replicated via this method in VMC. Landström and Granhag (2008) have also demonstrated that employing a long shot of a child in a mock eyewitness scenario is more beneficial than a close up shot, in so far as the witness was viewed in a more positive light. Sanford et al (2004) investigated the impact of half-duplex (low quality) and full-duplex (high quality) audio on task performance using the Map Task. Participants in the full-duplex condition interrupted one another more frequently and completed the task less accurately. Participants in the half-duplex condition however performed equally as well as face-to-face participants. It would seem that when using imperfect technology, the users adapted to the constraints of the technology, whereas those who used perfect audio did not make sufficient allowance for the VMC context, treating it like a face-to-face meeting. What these studies suggest is that users can adapt to the limitations of the technology by utilising simple techniques. These strategies are likely to lead to a more satisfactory exchange, ergo they lead to more favourable impressions.

CONCLUSION

In everyday life we commonly encounter situations in which we need to make judgements about others. Although this may not always be an entirely conscious process, the manner in which we respond to one another is critical for creating a rapport and establishing mutual understanding, which in turn provides impetus to continue with interactions (Marinetti et al, 2008). It is also clear that, for many people, the manner in which they present themselves to others is an important and frequent consideration. Decisions about what clothes we wear to an important job interview or how we phrase a crucial message for instance are decisive in putting across the 'right' image. Non-verbal cues are said to be particularly important in the impression formation process and others often judge us on our use of eye contact, gesture, posture and facial expressions for example. Although we may have a certain degree of control over how we present ourselves to others in a traditional face-to-face context, the social interaction landscape is constantly changing with the emergence of new technology and far less is understood about how we manage impressions in technology-mediated communication situations.

Throughout this review, evidence has been put forward to suggest that the use of video-mediated technologies can have a profound impact on the manner in which others perceive us. In many examples, it has been demonstrated that video-mediated impressions are less favourable than face-to-face impressions, and this is likely caused by a number of factors that make VMC unique. For example, non-verbal cues, which are particularly salient in impression formation, may be distorted, attenuated or missing from communication. Individuals may also feel more image conscious when using VMC as the technology creates the sensation of being observed or being monitored. Consequently, the way in which we behave in VMC contexts may be quite different to the way in which we would behave in a similar face-to-face encounter. Although video-mediated technologies are used for a variety of purposes and many of these are recreational in nature (for example, catching up with a friend), there are a number of applications in which important decisions are made about the competencies, personalities and dispositions of individuals. For example, VMC in a courtroom environment involves jury members passing judgment on witnesses, and this will not only be influenced by the evidence provided, but also the demeanour and behaviour of the witness. As the impression formation process is often an unconscious one, witnesses are unlikely to evaluate the significant contribution that the technology plays in the impression formation process.

Video-mediated impressions are influenced by a number of factors that are out of the control of interlocutors. However, although video-mediated impressions may be more negative, this is not say that they are less accurate and future research may wish consider the precision of judgements in cue restricted contexts. Further research may also evaluate many of the techniques that have been proposed in this review to improve video-mediated impressions. For example, the use of video-mediated gazing has been shown to improve participants' encoding and subsequent recall of information. Considering the important role played by eye gaze in person perceptions, it may therefore be worthwhile studying this technique in impression formation contexts. As the Internet and mobile communication devices continue to grow in popularity, the way in which we interact with others is also changing considerably. An understanding of the manner in which the use of technology-mediated communication influences how others perceive us would therefore seem more vital than ever before. With a deeper understanding of the ways in which technology interfere with our judgments of others, as well the techniques that may be employed to compensate for these problems, video-mediated communication may add greater value to remote communications.

REFERENCES

Abbott, L., Dallat, J., Livingston, R. & Robinson, A. (1993). The application of videoconferencing to the advancement of independent group learning for professional development. *ETTI, 31(2).*

Alexander, W., Higgison, C. & Mogey, N. (Eds). (1999). *Videoconferencing in teaching and learning: Case Studies*. LTDI and TALiSMAN, Institute for Computer Based Learning, Heriot-Watt University, Edinburgh.

Anderson, A., H. (2006). Achieving understanding in face-to-face and video-mediated multiparty interactions. *Discourse Processes, 41(3)*, 251-287.

Angiolillo, J. S., Blanchard, H. E., Israelski, E. W. & Mane, A. (1997). *Technology constraints of video-mediated communication*. In K. E., Finn, A. J. Sellen, & S. B. Wilbur, (Eds.). *Video mediated communication*. Lawrence Erlbaum Associates: New Jersey.

Argyle, M. (1994). *The psychology of interpersonal behaviour*. Penguin Books: London.

Argyle, M. & Dean, J. (1965). Eye contact, distance, and affiliation. *Sociometry, 28*, 289-304.

Ash, S. (1946). Forming impressions of personality. *Journal of Abnormal and Social Psychology, 41*, 258-290.

Bar, M., Neta, M. & Linz, H. (2006). Very first impressions. *Emotion, 6*, 269-278.

Beattie, G. W. (1981). The regulation of speaker turns in face-to-face conversation: Some implications for conversation. *Semiotica, 34*, 55-70.

Ben-Ze'ev, A. (2004). *Love online: Emotions on the Internet.* Cambridge University Press: Cambridge.

Bernieri, F. J. & Rosenthal, R. (1991) Coordinated movement in human interaction. In R. S. Feldman, & B. Rime, (Eds.) *Fundamentals of nonverbal behaviour*. Cambridge University Press: New York.

Brown, G., Anderson, A. H., Yule, G. & Shillcock, R. (1984). *Teaching talk*. Cambridge: Cambridge University Press.

Burns, J. A. & Kintz, B. L. (1976). Eye contact while lying during an interview. *Bulletin of the Psychometrics Society, 7*, 87-89.

Campbell, J. A. (1998). Participation in videoconferenced meetings: User disposition and meeting context. *Information and Management, 34*, 329-338.

Chapman, D. S. & Rowe, P. M. (2002). The influence of videoconference technology and interview structure on the recruiting function of the employment interview: A field experiment. *International Journal of Selection and Assessment, 10*, 185-197.

Chapman, D. S. & Webster, J. (2001). Rater correction processes in applicant selection using videoconference technology: The role of attributions. *Journal of Applied Social Psychology, 31(12)*, 2518-2537.

Chapman, D. S., Uggerslev, K. L. & Webster, J. (2003). Applicant reactions to face-to-face and technology-mediated interviews:. a field investigation. *Journal of Applied Psychology, 88(5)*, 944-953.

Clark, H. & Brennan, S. (1991). Grounding in communication. In L., Resnick, J, Levine, & S. Teasley, (Eds.) *Perspectives on socially shared cognition*, (127-149).

Colky, D. & Young, W. (2006). Mentoring in the virtual organisation: Keys to building successful schools and business. *Mentoring Tutoring: Partnership in Learning, 14*, 433-447.

Condon, W. S. (1982). Cultural micro rhythms. In M. Davis, (Ed.) *Interaction rhythms: Periodicity in communicative behaviour.* Human Sciences Press, New York.

Daft, R. L. & Lengel, R. H. (1986). A proposed interrogation among organisational information requirements, media richness and structural design. *Management Science, 32 (5)*, 544-571.

Davies, G. & Noon, E. (1991). *An evaluation of the live link for child witnesses.* Home Office Report.

De Greef, P. & Ijsselsteijn, W. A. (2001). Social presence in a home tele-application. *Cyber-Psychology and Behaviour, 4(2)*, 307-315.

Dennis, A. R. & Kinney, S. T. (1998). Testing media richness theory in the new media: The effects of cues, feedback, and task equivocality. *Information Systems Research, 9(3)*, 256-273.

Derks, D., Bos. & Grumbkow, J. (2007). Emoticons and social interaction on the internet: The importance of social context. *Computers in Human Behaviour, 23*, 842-849.

Derrer, N. M., Fullwood, C., Davis, S., Martino, O. & Morris, N. (2006). An initial face-to-face meeting improves person-perceptions of interviewees across a video link. In: P. T. McCabe, (Ed.) *Contemporary Ergonomics, 2006.* London: Taylor and Francis.

Doherty-Sneddon, G., Anderson, A. H., O'Malley, C., Langton, S., Garrod, S. & Bruce, V. (1997). Face-to-face and video-mediated communication: A comparison of dialogue structure and task performance. *Journal of Experimental Psychology, 3(2)*, 105-125.

Doherty-Sneddon, G. & McCauley, S. (2000). Influence of video-mediation on adult-child interviews: Implications for the use of the live link with child witnesses. *Applied Cognitive Psychology, 14*, 379-392.

Driskell, J. E. & Radtke, P. H. (2003). The effect of gesture on speech production and comprehension. *Human Factors, 45*, 445-451.

Driskell, J. E., Radtke, P. H. & Sales, E. (2003). Virtual teams: Effects of technological mediation on team performance. *Group Dynamics: Theory, Research and Practice, 7*, 297-323.

Ellsworth, P. C. (1975). Direct gazing as a social stimulus: The example of aggression. In P., Pliner, L. Krames, & T. Alloway, (Eds.). *Nonverbal communication of aggression*, (53-76). New York: Plenum Press.

Exline, R. V. (1971). Visual interaction: The glances of power and preference. In J. K. Cole, (Ed.). *Nebraska Symposium on Motivation*, (19, 162-205). Lincoln: University of Nebraska Press.

Ferran-Urdaneta, C. (2001). The effects of videoconferencing on persuasion. *Dissertation Abstracts International, 61(8-A)*, 3239, US: University Microfilms International.

Fullwood, C., Derrer, N. M., Martino, O. I., Davis, S. & Morris, N. (2006). The effect of an icebreaker on collaborative performance across a video link. In P. D. Bust, (Ed.) *Contemporary Ergonomics, 2006.* London: Taylor and Francis.

Fullwood, C. & Doherty-Sneddon, G. (2006) Effect of gazing at the camera during a video link on recall, *Applied Ergonomics, 37(2)*, 167-175.

Fullwood, C. (2007). The effect of mediation on impression formation: A comparison of face-to-face & video-mediated conditions. *Applied Ergonomics*, *38*, 267-273.

Fullwood, C., Judd, A. M. & Finn, M. (2008). The effect of initial meeting context and video-mediation on jury perceptions of an eyewitness. *Internet Journal of Criminology*, Primary research papers.

Gaver, W., Moran, T., Maclean, A., Lovstrand, L., Doursih, P., Carter, K. & Buxton, W. (1992). Realizing a video environment: EuroPARC's RAVE system. In P., Bauersfeld, J. Bennett, & G. Lynch, (Eds). *CHI' 92 Conference Proceedings*, (27-35), New York: ACM Press.

Goodman, G. S., Toby, A. E., Batterman-Faunce, J. M., Orcutt, H., Thomas, S., Shapiro, C. & Sashsenmaier, T. (1998). Face to face confrontation: Effects of closed circuit technology on children's eyewitness testimony and juror's decisions. *Law and Human Behaviour*, *22*, 165-203.

Grahe, J. E. & Bernieri, F. J. (1999). The importance of nonverbal cues in judging rapport. *Journal of Nonverbal Behaviour*, *23*, 253-269.

Haldigo-Barnes, M. & Massaro, D. (2007). Read my lips: An animated face helps communicate musical lyrics. *Psychomusicology*, *19*, 3-12.

Handy, C. (1995). Trust and the virtual organisation. *Harvard Business Review*, *73(3)*, 40-50.

Hambley, L., O'Neil, T. & Kline, T. (2007). Virtual team leadership: The effects of leadership style and communication medium on team interaction styles and outcomes. *Organisational Behaviour and Human Decision Processes*, *103*, 1-20.

Heath, C. & Luff, P. (1991). Disembodied conduct: Communication through video in a multimedia environment. *Proceedings of CHI '91 Human Factors in Computing Systems*, (99-103), New York: ACM.

Heath, C., Luff, P. & Sellen, A. (1995). From video-mediated communication to technologies for collaboration: re-configuring media space. In S. J. Emmott, (Ed.), *Information Superhighways: Multimedia Users and Futures*. Academic Press Ltd., London.

Hemsley, G. D. & Doob, A. N. (1978). The effect of looking behavior on perceptions of a communicator's credibility. *Journal of Applied Social Psychology*, *8*, 136-144.

Hinds, P. J. & Bailey, D. E. (2003). Out of sight, out of sync: understanding conflict in distributed teams. *Organization Science*, *14*, 615-632.

Horn, D. B. (2001). Is seeing believing? Detecting deception in technology mediated communication. *Proceedings of CHI 2001*, Short papers, New York: ACM press.

Huang, W. (2005). Social dynamics can be distorted in Video-Mediated Communication. *Dissertation Abstracts International Section A: Humanities and Social Sciences*, *66(2-A)*, 391.

Internet World stats, (2009). Retrieved from http://www.internetworldstats.com/stats.htm on 23rd July, 2009.

Isaacs, E. A. & Tang, J. C. (1993). What video can and can't do for collaboration: A case study. *Proceedings of ACM Multimedia 93*, (199-206), New York: ACM Press.

Ishii, H. & Kobayahsi, M. (1992). Clearboard: A seamless medium for shared drawing and conversation with eye contact. In *Proceedings of ACM CHI '92 Conference on Human Factors in Computing Systems*, (525-532), New York, ACM Press.

Kenyon, N. D., White, T. A. & Reid, G. M. (1985). Behavioral and user needs for teleconferencing. *Proceedings of the IEEE*, 73 (689-699). Reprinted in D.Bodson & R.Schaphorst (1989). *Teleconferencing*, (219-229), New York: IEEE.

Kiesler, S. & Cummings, J. (2002). What do we know about proximity and distance in work groups? A legacy of research. In P. Hinds, & S. Kiesler, (Eds.) *Distributed Work*, (57-80). Cambridge, MA: MIT Press.

Kiesler, S., Siegel, J. & McGuire, T. W. (1984). Social psychological aspects of computer-mediated communication. *American Psychologist*, *39*, 1123-1134.

Kitawaki, N., Kurita, T. & Itoh, K. (1991). Effects of delay on speech quality. *NTT Review*, *3*, 88-94.

Kleck, R. & Nuessle, W. (1968). Congruence between the indicative and communicative functions of eye contact in interpersonal relations. *British Journal of Social and Clinical Psychology*, *7*, 241-246.

Kraut, R. E. & Poe, D. (1980). Behavioral roots of person perception: The deception judgements of customs inspectors and laymen. *Journal of Personality and Social Psychology*, *39*, 784-798.

Landström, S., Granhag P. A. & Hartwig, M. (2005). Witnesses appearing live versus on video: Effects on observers' perception, veracity assessments and memory. *Applied Cognitive Psychology*, *19*, 913-933.

Landström, S., Granhag, P. A. & Hartwig, M. (2007). Children's live and videotaped testimonies: How presentation mode affects observers' perception, assessment and memory. *Legal and Criminological Psychology*, *12*, 333-347.

Landström, S. & Granhag, P. A. (2008). Children's truthful and deceptive testimonies: How camera perspective affects adult observers' perception and assessment. *Psychology, Crime and Law*, *14(5)*, 381-396.

Larsen, R. J. & Shackelford, T. K. (1996). Gaze avoidance: personality and social judgments of people who avoid direct face-to-face contact. *Personality and Individual Differences*, *21(6)*, 907-917.

Marinetti, C., Moore, P., dos Anjos, P. L. & & Parkison, B. (2008). Emotions in social interactions: Construction of emotion experience. Retrieved from http://cfpm.org/~pablo/anjos,humaine_chapter.pdf on August 19th, 2009.

Martin, C. L. & Nagao, D. H. (1989). Some effects of computerized interviewing on job applicant responses. *Journal of Applied Psychology*, *74(1)*, 310-314.

Martino, O. I., Fullwood, C., Davis, S. J., Derrer, N. M. & Morris, N. (2006). Maximising VMC: The effects of system quality and set-up on communicative success. In P. D. Bust, (Ed.) *Contemporary Ergonomics*, *Taylor and Francis*, 139-143.

McCulloch, M. S. & Oborne, D. J. (1999). Video conferencing systems: Telepresence and selection interviews. In M. A., Hanson, E. J. Lovesey, & S. A Robertson, (Eds.) *Contemporary Ergonomics*, 133-137.

Mehrabrian, A., & Williams, M. (1969). Nonverbal concomitants of perceived and intended persuasiveness. *Journal of Personality and Social Psychology*, *13*, 37-58.

Morley, I. E. & Stephenson, G. M. (1969). Interpersonal and interparty exchange: A laboratory simulation of an industrial negotiation at the plant level. *British Journal of Psychology*, *60*, 543-545.

O'Conaill, B., Whittaker, S. & Wilbur, S. (1993). Conversations over videoconferences: An evaluation of the spoken aspects of video mediated communication, *Human-Computer Interaction*, *8*, 389-428.

O'Malley, C., Langdon, S., Anderson, A. & Doherty-Sneddon, G. (1996). Comparison of face-to-face and video mediated interaction. *Interacting with computers*, *8*, 939-941.

Orcutt, H. (1995). *Detecting deception: Fact finders' abilities to assess the truth*, Doctoral dissertation, State University of new York at Buffalo, NY.

Parkinson, B. (2008). Emotions in direct and remote social interaction: Getting through the spaces between us. *Computers in Human Behaviour*, *24(4)*, 1510-1529.

Parkinson, B., Fischer, A. H. & Manstead, A. S. R. (2005) *Emotion in social relations: Cultural, group and interpersonal processes*. Psychology Press, New York.

Parkinson, B. & Lea, M. (in press). Video-linking emotions. In A. Kappas, (Ed.). *Face to face communication over the Internet: Issues, research, challenges*. Cambridge: Cambridge University Press.

Pellegrini, R. J., Hicks, R. A. & Gordon, L. (1970). The effects of an approval-seeking induction on eye-contact in dyads. *British Journal of Social and Clinical Psychology*, *9*, 373-374.

Rainie, L. (2005). Use of web cams. Pew American and Internet life project reports. Retrieved from http://www.pewinternet.org/Reports/2005/Use-of-web-cams.aspx on July 23rd, 2009.

Rice, R. E. (1993). Media appropriateness: Using social presence theory to compare traditional and new organizational media. *Human Communication*, *19*, 451-484.

Rocco, E. (1998). Trust breaks down in electronic contexts but can be repaired by some initial face to face contacts. *Proceedings of CHI*, 1998, 496-502.

Rockman, K. W. & Northcraft, G. B. (2008). To be or not to be trusted: The influence of media richness on defection and deception. *Organisational Behaviour and Human Decision Processes*, *107(2)*, 106-122.

Rutter, D. (1987). *Communicating by Telephone*. Pergamon.

Sanford, A., Anderson, A. H. & Mullin, J. (2004). Audio channel constraints in video-mediated communication. *Interacting with Computers*, *16(6)*, 1069-1094.

Scottish Office: Central research unit, Crime and criminal justice research findings No. 4. (1995). Live television: *An evaluation of its use by child witnesses in Scottish criminal trials*. The Scottish Office.

Sellen, A. (1997). Assessing video mediated conduct: A discussion of different analytic approaches. In K.E. Finn, A. J. Sellen, & S. B. Wilbur, (Eds). *Video Mediated Communication*. Lawrence Erlbaum Associates, New Jersey..

Sproull, L. & Kiesler, S. (1986) Reducing social context cues: Electronic mail in organised communication. *Management Science*, *32*, 1492-1512.

Storck, J. & Sproull, L. (1995). Through a glass darkly. What do people learn in videoconferences? *Human Communication Research*, *22(2)*, 197-219.

Straus, S. G., Miles, J. A. & Levesque, L. L. (2001). The effects of videoconference, telephone, and face-to-face media on interviewer and applicant judgments in employment interviews. *Journal of Management*, *27(3)*, 363-381.

Strom, D. (1997). Videoconferencing in Focus. *Internet World Magazine, 8(9)*, Mecklermedia Corporation.

Suler, J. R. (2004). The online disinhibiton effect. *Cyber Psychology and Behavior, 7*, 321-326.

Sylvester, J., Anderson, N., Haddleton, E. Cunningham-Snell, N. & Gibb, A. (2000). A cross-modal comparison of telephone and face to face selection interviews in graduate recruitment. *International Journal of Selection and Assessment, 8*, 16-21.

Tankard, J.W. (1970). Effects of eye position on person perception, *Perceptual and Motor Skills, 31*, 883-93.

Taylor, N. & Joudo, J. (2005). The impact of pre-recorded video and closed circuit television testimony by adult sexual assault complainants on jury decision making: an experimental study,' *Research and Public Policy Series, Australian Institute of Criminology.*

Tiffin, J. W. & Rajasingham, L. (1995). *In search of the virtual class. Education in an Information Society*. Chapter 6: Telelearning in Cyberspace.

Videotalk. (1998). Retrieved from http://www.videotalk.co.uk/home.html on February 17th, 1999. © 1998 Wincroft Inc.

Watson, A. & Sasse, M. A. (1996). Evaluating audio and video quality in low cost multimedia conferencing systems. *Interacting with Computers, 8(3)*, 255-275.

Wegge, J. (2006). Communication via videoconference: Emotional and cognitive consequences of affective personality dispositions, seeing one's own picture and disturbing events. *Human-Computer Interaction, 21(3)*, 273-318.

Whittaker, S. (1995). Video as a technology for interpersonal communication: A new perspective. *IS&T SPIE Symposium on electronic imaging science and technology, 2417*, 294-304.

Whittaker, S. (1996). Rethinking video as a technology for interpersonal communication: Theory and design implications. *International Journal of Human Computer Studies, 42*, 501-529.

Williams, E. (1977). Experimental comparisons of face-to-face and mediated communication: A review. *Psychological Bulletin, 84(5)*, 963-976.

Yerkes, R. M. & Dodson, J. D. (1908). The relation of strength of stimulus to rapidity of habit-formation. *Journal of Comparative Neurology and Psychology, 18*, 459-482.

In: Videoconferencing: Technology, Impact and Applications ISBN: 978-1-61668-285-9
Editor: Adam C. Rayler, pp.57-79

Chapter 3

APPLICATIONS OF VIDEOCONFERENCING TECHNOLOGIES IN BIOENGINEERING

Carlos Dafonte, Ángel Gómez, Diego Ordóñez and Bernardino Arcay
Department of Information and Communications Technologies,
University of A Coruña, 15071 A Coruña, Spain

ABSTRACT

Technological advances have extended the use of videoconferencing systems in a variety of areas, such as education, health and medicine, business, law, security, private use, etc. The health and medicine area in particular strives at improving clinical attention to patients by implanting videoconferencing into specific devices (equipped rooms) and general devices (systems installed in personal devices) so as to improve the clinical attention that is given to patients. Videoconferencing is implanted in telemedicine, teleassistance, telesurgery, etc. in order to help sanitary staff with the elaboration of diagnoses, assistance to patients in rural areas, the interactive exchange of clinical information between specialists, assistance in long-distance operations, etc.

Videoconferencing requires the presence of an equipped room or a device (PC, PDA, mobile phone, etc.) that contains the necessary hardware to support it (video camera, microphone and speaker, specific software). The videoconferencing systems that are used in the medical/sanitary field are similar to those of other environments, except for the presence of peripheral devices such as medical images devices, microscopes, monitoring devices, etc, which may require additional signal transmission.

In many cases, maintaining a certain quality of image and sound in real time is essential, and therefore very high bandwidth becomes indispensable. For this reason, videoconferencing systems use compression formats that can be adapted to the existing transmission means while maintaining an acceptable quality of image and sound (service quality). Also, these systems must follow audio/video standards that make it possible to implant them into various systems and adapt them to specific tasks.

The videoconferencing standards that are used in the fields of health and medicine are mainly the following: H.320, a standard that is designed for ISDN communications and groups the H.261 norm for video, G.711/G.722/G.728 for audio, and T.120 for data;

and H.323, a standard that is designed for IP communications and groups the H.261/H.323 norms for video, G.711/G.722/G.723.1/G.728 for audio, and T.120 for data.

INTRODUCTION

The medical/sanitary field is of great importance and requires the planning of health policies that improve the provided assistance and increase the productivity of sanitary practices while reducing the associated costs.

The latest advances in information and telecommunications technologies, as well as their cost reduction, have opened up the future of bioengineering in medicine and healthcare. Many specialties are now equipped with Telemedicine/Telehealth applications that provide a new, practical, and effective way of offering more people less expensive and time-consuming medicine by considerably optimizing resources.

Applications of this type can be grouped into three categories, according to their functions:

- **Clinical applications.** This is the largest group and includes, for example, applications for specialized medical devices, diagnostics, etc. It can be subdivided into clearly differentiated development areas:
  - **Diagnostic images.** This group contains a series of applications for the management of radiological images, images of pathological anatomy and dermatology, and other images with clinical relevance, as well as for their analysis, storage, and integration with other medical applications and their exchange through network systems. The PACS (Picture Archiving and Communication Systems) correspond to the applications that are proper of this group and are very useful in radiology.
  - **Laboratory examinations.** In this group, systems are developed that allow the integration of information devices for administrative information as well as examination results, quality control, process automatization, etc., and their communication with other health information systems (such as the hospital information system, the clinical file, etc.).
  - **Patient monitoring.** This group is dedicated to the registration of patient information that directly refers to the daily evolution of vital signals, infirmary registers, and electronic monitoring (ECG, mechanical ventilation, oxymetry, infusions, sleep monitoring, etc.).
- **Administrative applications.** Applications that are focused on the administration of medical cares such as Hospital Information Systems (HIS), in charge of the administrative, demographical, financial, and administrative patient information.

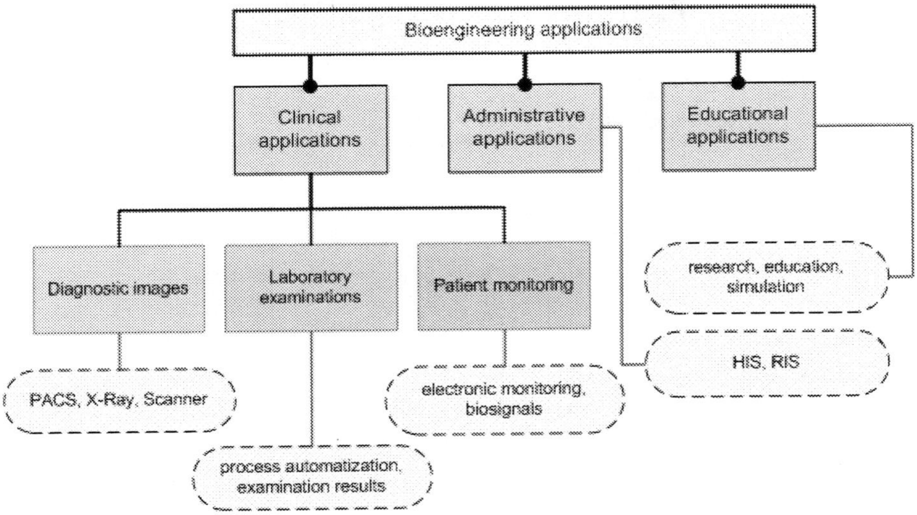

Figure 1. Bioengineering applications

- **Educational applications.** This group includes research and teaching applications and services (education) and allows for the educational upgrade of the medical staff, the exchange of opinions on a clinical file, etc. We can mention the following examples: expert systems for support to diagnostics (e.g. Dxplain, a programme that generates differential diagnoses for teaching and clinical application), medical databases (e.g. MEDLINE), Unified Medical Language System (UMLS), multimedia and simulation programmes, etc.

To a large extent, these systems are completed by a module that provides videoconferencing for daily practice and allows real-time communication through audio and video between the sanitary staff and the patient, or between medical staff located in different places.

TELEMEDICINE AND TELEHEALTH

There are many ways to define "telemedicine", but they all reflect the same concept: the application of information technologies to the provision of medical services and/or data and their transfer from one place to another. A more concrete definition could be the following:

"Telemedicine is the use of advanced telecommunication technologies in the exchange of health information and the provision of healthcare, regardless of geographical, socio-cultural and temporal barriers."

In 1975, Bird defined telemedicine as follows:

"Telemedicine is the practice of medicine without the usual physician-patient physical confrontation via an interactive audio-video communication system" [Bird, 1975].

In 1983, Conrath et al, proposed a more general definition:

"Telemedicine is the use of communications technologies to assist in the delivery of health care" [Conrath, 1983].

Medical informatics constitutes an important part of telemedicine because it reflects the application of data technology to data storage, systemization, management, and refinement within medical institutions. Such applications can be for administrative or purely medical purposes.

In many contexts, the concept of telemedicine is reserved for applications whose purpose consists in providing health care based on telecommunications applications, such as remote consultations and diagnostics in various medical specialties. In some cases, the transmission of knowledge in the shape of distance education and remote instructions is included in the concept of telemedicine. As such, we can find the following definition:

"The research, monitoring, and management of patients and the education of patients and employees by means of systems that provide direct access to advice from experts and to patient information, regardless of where the patient or the relevant information are located" [AIM, 1990].

According to the World Health Organization, the terms Telemedicine/Telehealth are part of the definition of Informatics in Health, since they include telematics:

"Informatics in Health is a broad concept that embraces the rapidly evolving disciplines of computation, network creation, and communications (methodology and technology) which provide support to areas linked to health, such as Medicine, Nursing, Pharmacy, and Dentistry" [WHO, 1988].

Finally, another equally valid definition could be the following:

"Telemedicine is the use of electronic information and communication technologies to provide and support medical care when distance separates the involved participants" [IM, 1996].

Perform a medical consultation through videoconferencing and without the physical presence of the patient, access the clinical files of a patient through a computer located in a remote sanitary center, operate on a patient located many miles away,...long-distance medicine provides, or will provide in the future, these and many more possibilities.

Many examples illustrate the existing need for the use of telemedicine techniques. Companies that relocate their employees to isolated areas with scarce sanitary resources (construction companies, oil companies, etc.) or to rural areas far removed from large urban areas are known to have implemented telemedicine systems in order to have fast and efficient access to the most modern medical technology under the supervision and support of experts in each field. Another example is military field medicine, where modern armies consider these types of systems as a basic part of their military and sanitary infrastructure.

The technology that is presently at hand uses data networks and digitalized radio signals to transmit signals that can be channeled through satellites and to receive high quality compressed audiovisual images such as X-Rays, scanners, video ultrasounds, electrocardiograms, or even a real long-distance consultation between medical experts; it allows the transmission of medical details that go from cardiopulmonary auscultation to the percussion of an organic cavity. The use of new technologies (multimedia systems, artificial vision, neural networks, nanominiaturization, non-invasive techniques) associated to telecommunications will determine the future of medicine. The worldwide generalization of the Internet in particular will facilitate the implementation of large amounts of services on top of the already existing ones. For instance, it will allow the gathering of information from universities, libraries, medical publications, medical centers, and specialized users in a simple and fast manner. Interestingly, it will also allow the obtention of a "second opinion" by confronting experts in concrete fields through videoconferencing, this being one of the most interesting and increasingly preponderant aspects of telemedicine. Ordinary medicine, rural medicine, sports medicine, emergency medicine (catastrophes, earthquakes, floods, etc.),…any type of medicine that requires experts in concrete cases can find an excellent means of communication and team work.

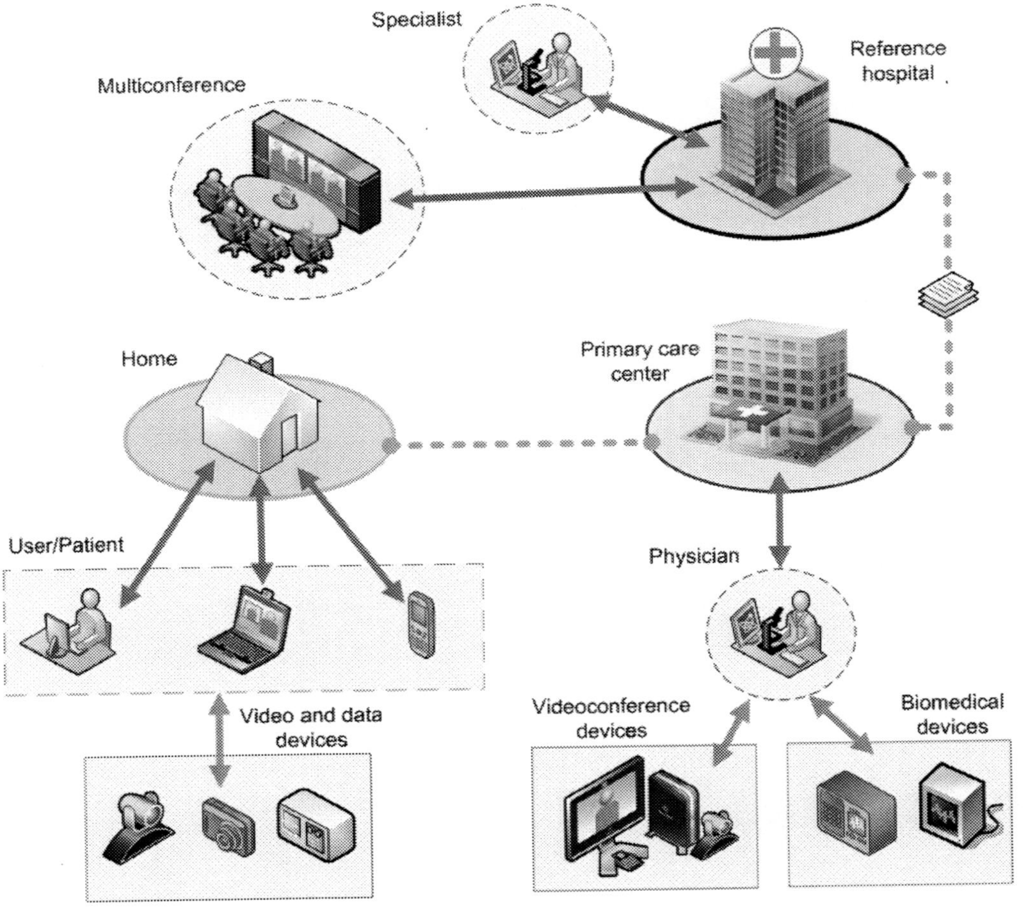

Figure 2. Telemedicine operating diagram

Within telemedicine, the use of artificial intelligence techniques and in particular expert systems makes it possible to solve problems in a very concrete field of experience, such as diagnostics, epidemic prevention, disease development simulation, treatment planning, and graphic testing interpretation. More concretely, telediagnostics allow the medical expert to consult a case with a center through the network and receive a response that might be from a specialist or an expert system; in both cases the purpose of artificial intelligence is to help the expert by saving him routine work but leaving the final decision in his hands.

The main advantage of telemedicine lies with the fact that the physical presence of the medical expert is not required. This avoids any unnecessary relocation of patients from rural areas and allows sanitary attention to reach isolated centers in a faster and more specialized manner. Also, it reduces collapses and waiting lists in large hospitals and guarantees patient attention in spite of the decentralization of medical specialties. Even so, experts admit that long-distance medicine raises several questions: who is responsible for a consensuated computer diagnosis; what authority is to be given to foreign specialists, who are just as accessible as local specialists; how is the free election of doctors to be regulated in order to avoid that only the best specialists are consulted; how to compensate doctors for their additional work; how to control the access the confidential clinical information; and how to make hospitals organize themselves quickly without fear of increasing their present workload?

Telemedicine provides, among others, the following services:

- **Diagnostics:** digitalization allows medical experts to send the patient data from a medical ward to the most indicated specialist. At the present, the most developed technique in this field is teleradiology. Another application of remote diagnostics is telesonography or tele-ultrasounds, based on the transfer of video and patient data in real time from an ultrasound device.

- **Videoconferencing:** videophones add images to telephone communication and are calculated to reduce actual visits to the doctor by half. Videoconferencing is used for routine consultations, follow-up of pregnancies, accidents, treatments, etc, and allows the specialist to interview the patient or his/her generalist doctor. In some cases, high image and sound quality are required for medical practice, which implies the availability of large bandwidth and adequate devices.

- **Common clinical files:** the sanitary centers automatically access the medical files that are archived elsewhere.

- **Internet:** at home, the patient can consult the guides for clinical practice and find more information on the medical center that he/she will finally opt for (methods, experiences, tariffs, specialists, etc.).

- **Robot-assisted surgery:** surgical operations are carried out by means of simulators that use virtual reality techniques and a complex high-quality system for audio and video that is able to control a robot system that reproduces the movements of the surgeon's hands. This technique is especially indicated for maximum precision operations.

- **Mobile devices:** the use of small mobile devices (PDA, UMPC, mobile telephone, etc.) allows the expert to access information and take notes by dictating or writing; the information is automatically stored in the expert's workstation so as to avoid any loss of data.

- **Intelligent monitoring:** an artificial intelligence expert system registers patient variables in real time, uses certain parameters to warn in case of alarm signals or risk prognoses, and provides immediate assistance.

Finally, and in spite of the tremendous medical and economical advances that the progressive development of telemedicine may entail, we must bear in mind that the medical act as such, the act that links the patient to its doctor in a very personal manner, is and will always remain an irreplaceable act.

Standards in Medical Informatics

In order to unify the development of computerized solutions in medical computer sciences, researchers elaborated a series of norms or standards that allow the integration of computers and applications for biomedical use. This obliged the manufacturers of hardware and software to comply with certain standards that facilitate the intercommunication between systems:

- DICOM. A standard that normalizes the exchange of clinical images. In 1985, the *American College of Radiology* (ACR) and the *National Electrical Manufacturers Association* (NEMA) published a standard for the transmission of relevant data in medical images called DICOM (*Digital Imaging and Communications in Medicine*). A revised version was published in 1988. In both versions, the data transmission was defined by means of point-to-point connections, without considering a network environment (LAN or WAN). The latest versions of this norm were developed so as to comply with the requirements of manufacturers and users with regard to the connection of medical images devices to standard networks. The norm is not only in charge of the storage, impression, and transmission of images that proceed from medical devices, it also includes the exchange of reports and video recordings in MPEG-2 format.
- HL7. A set of standards for the exchange of data between different health information systems. HL7 (*Health Level-7*) was created in 1987 to develop standards for the electronic exchange of clinical, administrative, and financial information between independent computers oriented towards health care (e.g. hospital information systems, clinical laboratory systems, pharmaceutical systems, etc.). This standard is placed at the highest level of the OSI model levels (*Open System Interconnection*) so as to allow the interpretation and communication of data between

different computer applications and architectures. In June 1994, the *American National Standards Institute* (ANSI) opted for HL7 as an accredited standard developer.

- ASTM. A standard that was developed for the exchange of messages related to clinical and electrophysiological observations and medical logics. Founded in 1898, the *American Society for Testing and Materials* (ASTM) is a non-profit organization that provides a forum for producers, users, and other interested parties aimed at creating a consensus and designing standards for materials, products, systems, and services. The E-31 committee is in charge of the standards for the interconnection of laboratory equipment, computers, bar codes, medical records and their confidentiality and privacy policies, authentification, networks, etc.
- IEEE. Development of communication standards between medical devices and information networks. The *Institute of Electrical and Electronics Engineers* (IEEE) is the largest professional technical society in the world. Its technical purposes focus on advancing the theory and practice of computer engineering and computational science. In the medical area, this organization has worked on developing standards for the interconnection of medical equipment, modelling of medical information, etc. There are two main IEEE standards on healthcare:

  IEEE 1073, a standard for medical communication devices that consists in a series of documents that define the requirements for the seven communication layers of the "*Medical Information Bus*" (MIB). It was designed to provide a robust and reliable communication service for Intensive Care Units, operating rooms, and bedside devices in emergency rooms.

  IEEE 1157, a norm for the exchange of medical data that consists in a series of documents that define the communication model for the exchange of medical data between various systems (this norm is also known as "MEDIX").
- CCO. Development of standards for the coordination of multiple applications in a workstation. The *Clinical Context Object Workgroup* (CCO) is a group of salespeople and users that specified the interfaces as a standard component that could serve as context coordinator. The involved applications work together by sharing patient information and other types of data, so that when the information of one application is changed, all the applications are updated.
- CEN/TC 251. A European initiative for the establishment of standards in medical informatics and telematics. In 1990, the *European Committee for Standardisation* (ECN) created this technical committee for computers sciences applied to health. The purposes of CEN/TC 251 are the organization, coordination, and follow-up of standard developement, including the testing of standards in medical informatics and telematics at a European level. The completed CEN/TC 251 standards are destined to the health care sector.

Communications Technologies for Videoconferencing

Since telemedicine is mostly aimed at the transmission of medical data between rural and urban areas, its technology should make the most of the existing communication infrastructure in order to be cost-efficient. Copper telephone lines are the most common communication channel between distant communicators, but unfortunately they were not designed for rapid digital communication and their possibilities as to how data can be transferred are rather limited. This means that the videoconferencing communications will have to be adapted to the available bandwidth without jeopardizing the clinical practice. In some cases, the transfer of medical data will have priority over video communication, in other cases the video quality will be of less importance and a larger compression factor will be applied.

The already classical modem-based systems for data communication over telephone lines provide the transfer of data with low bandwidth (typical modem speeds reach 14,4 kb/s, 28,8 kb/s, or 56 kb/s) and, therefore, with low service quality.

ISDN slightly improved the potential bandwidth for digital data transfer thanks to the use of telephone switches and already existing wiring, and hereby allowed for important advances in telemedicine applications. Basic ISDN provides 64 kb/s for data transmission and two voice lines. Also, more ISDN lines can be added so as to provide more bandwidth.

In recent years, the introduction of new data transmission technologies such as DSL (*Digital Subscriber Line*) have increased the bandwidth for cable communications, hereby offering an immediate and cost-efficient means to connect to the Internet and to remote access services for end-users. DSL can be subdivided into different modalities, among which are ADSL (*Asymmetric Digital Subscriber Line*) and standard ANSI T1.413, which provide high capacity data transmission over currently existing telephone lines and applications such as videoconferencing. Among the "G-Series" of the ITU-T are the recommendations concerning the different approved ADSL versions (G.992.1, G.992.2, G.992.3, G.992.4, and G.992.5 norms) as well as other DSL transmission norms.

By replacing the most common means of transmission, i.c. telephone lines, by other means such as coaxial cable and optical fibre, we can increase the transmission bandwidth, but we also increase expenses. Optical fibre connections provide speeds of over 150 Mb/s and can be made to reach the home of the users (FTTH - *Fiber To The Home*): this combined with transmission technologies such as ADSL results in very high bandwidth levels which are perfectly apt for telemedicine applications that require high-quality videoconferencing capacities.

TELEMEDICINE APPLICATIONS

Telemedicine embraces a wide range of applications, depending on the medical specialty that we wish to assist and the requirements of the communications, which can be either synchronous (the interaction takes place in real time) or asynchronous (the interaction does not take place in real time).

Asynchronous telemedicine systems mainly use medical data that do not require immediate interaction (e.g. medical images, X-Rays, computed axial tomographies, ultrasounds, biosignals, electronic medical files, etc.), even though they may include videoconferencing as a support means. Typical specialties for this kind of communication are dermatology, radiology, and pathology.

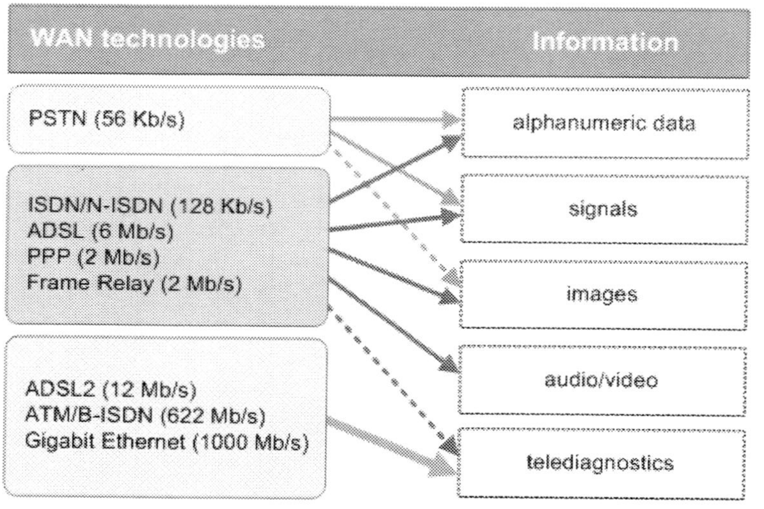

Figure 3. WAN technologies and Telemedicine capacities

In synchronous telemedicine systems, real-time interaction is particularly important, through videoconferencing as well as through medical devices connected to the system (e.g. tele-otoscopes for remote hearing examination, telestethoscopes for remote auscultation of heartbeat or respiration, tele-ultrasounds for the follow-up of pregnancies and the detection of traumas, tumors, etc.). Typical specialties for this type of communication are psychiatry, general medicine, internal medicine, rehabilitation, pediatrics, obstetrics, gynaecology, neurology, speech pathologies, and pharmacy.

The most noticeable applications in bioengineering are the following:

Teleconsulting Systems

These systems are aimed at saving time and trouble to users who normally have to travel to meet their medical expert: diagnostics and treatment are carried out by a specialist who refers to another specialized center or to the patient's home. In these applications, the videoconferencing communication module becomes particularly important for diagnostic purposes, which often require images and sounds of the highest quality. Based on the participants of teleconsultation, we can distinguish the following communication types:

- Doctor-patient communication, which facilitates the follow-up of the patient from the hospital. Examples: post-operatory follow-up, house calls, elderly people, pregnancy follow-up, medical checkups, first aid, etc.
- Generalist doctor-specialist communication, which facilitates the transmission of medical data, consultation and collaboration between experts, and the elaboration of diagnoses and prognoses.
- Patient-expert system communication: even though the final decisions are made by the medical staff, the expert system can elaborate conclusions on the basis of symptoms.

NORTH, the network from north Ontario, Canada, was launched in 1998 and has allowed users to carry out thousands of teleconsultations in more than 30 different specialties, especially in dermatology, psychiatry, pediatrics, movement disorders, and chronic pain. As a part of that network, the Hospital for Sick Children has performed almost 1500 teleconsultations in Pediatrics and Pediatric Surgery between 1996 and 2002, mostly related to post-operatory follow-up.

In the area of infantile medicine, the National Laboratory for Rural Telemedicine (Iowa, USA) performs teleconsultations related to infantile echocardiography. In a clinical test, this entity demonstrates the absence of significant differences between presential consultations and teleconsultations through videoconferencing, and it proposes its application to the prison environment (implanting mostly systems oriented towards psychiatry and dermatology).

One of the earliest telemedicine services was **teledermatology**, which today continues to be among the most common applications in telemedicine, providing consultations, diagnoses, and treatments of all types of skin-related illnesses and injuries, using both data and audio/video in real time. In certain cases, an interactive videoconferencing communication is not even necessary because the specialist disposes of the clinical patient files and the digital images of the affected area. The expression "teledermatology" was first used in 1995 to describe the value of a similar service in rural areas that did not have direct access to expert dermatologists [Perednia, 1995]. Nowadays we find multiple examples of teledermatology systems that use real-time videoconferencing modules [Loane, 1998] [Mars, 2008][Schmid-Grendelmeier, 2003].

Using videoconferencing as a central element, **telepsychiatry** has demonstrated its usefulness to the extent that it was accepted by the American Psychiatric Association in 1998. Penitentiary institutions were able to establish communication with external medical experts [Manguno-Mire, 2007][Manfredi, 2005], and rural areas could finally provide therapeutic services [Cowain, 2001][Hilty, 2006].

The European project EuroPaNet is yet another example in this category. EuroPaNet is aimed at providing psycho-pedagogical treatment to mentally disabled patients through videoconferencing, real-time interaction, and improved data analysis. The client-server applications facilitate parallel multi-user sessions of psycho-pedagogical and medical assistance [EuroPaNet, 2001][Kiekens, 2002].

Tele-Nursing and Tele-Assistance

This terminology refers to sanitary assistance through the home monitoring of chronic patients with diabetes, asthma, etc. More recently, applications have been developed that allow the follow-up of these illnesses through mobile telephones and specific medical devices [Istepanian, 2005]. One example of tele-assistance is ET-ASSIST (HC-1016) [ET-ASSIST, 1999], a European project whose purpose is to develop the collaboration between healthcare professionals involved in medical assistance, in order to improve the continuity and quality of health care across borders and in remote areas.

Mobile Centers/Tele-Ambulances

This terminology refers to the availability of high-quality health services in remote places thanks to the presence of special mobile units. These units can be moved and used for teleconsultations and the transmission of medical data to hospitals; they also allow the patient to be monitored and his/her data to be sent to the hospital in the course of a transfer. As an example we can mention the EU project XMOTION (eMobile Testbed For Interoperability Of Net-works In eLogistics), where mobile network technologies (2.5, 3G) are used to provide high bandwidth services for the management of emergencies, transport follow-up, and tele-ambulances [XMOTION, 2001].

Telecontrol Systems in Telesurgery

This terminology refers to the remote handling of instruments by means of a computer network. Telesurgery is increasingly applied in long-distance operations where robots replace the specialists [Guthart, 2000][Melvin, 2002].

Telesurgery became a possibility with the advent of laparoscopic surgery in the late 1980s. Laparoscopy is a surgical procedure in which a laparoscope (a thin lighted tube) and other instruments are inserted into the abdomen through small incisions. The internal operating field may then be visualized on a video monitor connected to the scope. In certain cases, the technique may be used in place of more invasive surgical procedures that require more extensive incisions and longer recovery times. Computer-assisted surgery premiered in the mid-1990s; it was the next step toward the goal of remote surgery. The computer-assisted system named *ZEUS Surgical System*, developed in 1995, was approved by the Federal Drug Administration (FDA) in 2002 for use in general and laparoscopic surgeries, but it nevertheless required the presence of both patient and surgeon in the same room.

Real, authentic telesurgery would require the collaboration of the communication industry to meet the needs for secure and reliable high-speed data transmission with- and this was fundamental-very little delay.

In 2000, the same company that developed the ZEUS system (*Computer Motion*), Jacques Marescaux (head of the European Institute of Telesurgery), and the French telecommunications company France Telecom started the *Operation Lindbergh* project. The first prototype, ZEUS TS, was finished near the end of 2000 and tested in France with distances of approximately 1000 Km. In 2001, testing was carried out in a clinical environment in New York City (USA) and Strasbourg (France), resulting in the first complete remote surgery on a human patient (a 68-year-old female), performed over a distance of 7000 km. The patient left the hospital within 48 hours, as is usual when using the laparoscopy cholecystectomy technique, and enjoyed full recovery.

In 2003 took place the first telerobotic-assisted inter-hospital surgery in Ontario (Canada). Two surgeons (one locally and the other at 400 Km distance) worked together in a *Nissen dundoplication* (surgery to correct chronic acid reflux).

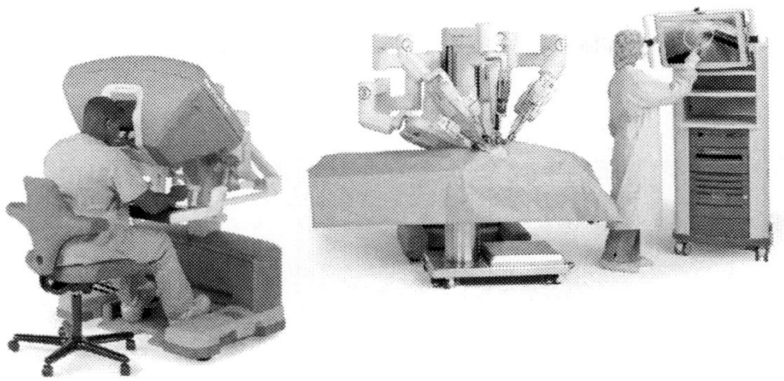

Figure 4. The *da Vinci* Surgical System (teleoperative machine designed by the company *Intuitive Surgical*)

The above experiences indicate that other potential applications could be the following:

- Training of inexperienced surgeons
- Assistance in development countries and rural areas with small populations
- Surgery in military areas
- Medical attention in space
- World-wide medical collaboration between surgeons
- Etc.

Teleradiology

This terminology refers to the exchange of reports and consultations based on the obtention of radiological images such as X-Rays, CTs (Computed Tomography), or MRIs (Magnetic Resonance Imaging), and their digital transmission to a radiologist in a remote hospital. Current radiological scanners include direct digitalization and storage of the images in normalized formats.

The development of teleradiology systems started in the 1990s and has led to the present systems [Thrall, 2002][Thrall, 2005]. Specifically interesting is the R-Bay project (http://www.r-bay.org), an EU-funded market validation project under the eTEN programme [eTEN, 2002] that aims at establishing an online eMarketplace within the field of radiology.

Telemonitoring Environments

This terminology refers to systems that allow the efficient real-time management of an intelligent monitoring system (e.g. Intensive Care Units, home care, dialysis, etc.) through the network of a hospital environment. These systems use sensors that capture and transmit biometric data of the patient for its subsequent visualization and treatment.

The next section briefly summarizes the architecture of the intelligent telemonitoring system for Intensive Care Units that was developed by the authors of this chapter. The implantation of this system required the integration of acquisition techniques, intelligent agents, and network traffic management [Dafonte, 2007].

Telemedicine Applied to Critical Telemonitoring

The mechanism consists in a real-time monitoring system for critical care units (CCU) that can function inside a hospital (LAN) as well as between hospitals (WAN with point-to-point or Internet connections).

The problems they are trying to solve are not trivial: for each patient in a critical care unit, physicians manage up to 30 variables that proceed from approximately 20 devices that monitor the patient's physiological variables [Eichhorn, 1986], from the results of laboratory tests, and from the parameters associated to vital support instrumentation such as mechanical ventilators or infusion pumps.

The design of monitoring and supervision systems in a CCU must consider all the following requirements:

- Integrate the biomedical monitoring instrumentation and facilitate the acquisition of data from other sources, such as laboratory analyses.
- Store the information and facilitate its recuperation: two processes for which databases are indispensable, since the design of the system largely depends on them.
- Dispose of ergonomical interfaces that allow the integration and exploitation of the data, the study of historic data, and the management and configuration of the instrumentation.
- Provide a support that allows the configuration and management of the alarm levels associated with the patient´s instrumentation.

For each interconnection between a bedside PC (BCC) and the central system (MCC) we establish the following channels (with the corresponding execution threads at each side):

- Digitalised Information Transmission (**DIT**). This transmission thread allows the exchange of files (images, documents, etc).
- Signals digitalised in Real Time (**SRT**). The signals are transmitted to the MCC for each acquisition of an intermediate storage or buffer in the BCC. Approximately 70% of the generated traffic is reduced by the data compression functions of the *zLib* library.
- Medical analysis and Control Information (**MCI**). This transmission thread allows us to transmit and receive the information from the central system concerning the remote control of the BCC (changes in the configuration of normality ranges in medical variables, analogous acquisition parameters, etc.). It is also in charge of sending the information generated by the intelligent module of the BCC: off-line data (e.g. a Glasgow Test), messaging, partial status of the physiological subsystems, stability analyses, and alarms.
- Videoconferencing system (**Videoconf**). The system is equipped with an application that was developed with the VDK of VCON®, and with VCON® Escort25 hardware,

compatible with the H.323 standards for videoconferencing and collaborative applications. We opted for this commercial solution because it allows us to dynamically modify the bandwidth consumption (which logically affects the provided quality).

A CCU monitoring system is not only in charge of a variety of tasks: signal and image processing, stability calculations, symbolic classification, alarms management,...; it must also perform these tasks in real time. This is why we thought of developing a 'Task Manager' that would be able to use the computational resources that exist in the network and distribute the tasks in it. This conception has a totally distributed character: it uses networked intelligent agents, which in turn include an intelligent module developed in OPS/R2 language.

On the other hand, in the course of the migration of the networked monitoring system towards a Telemedicine environment, we observed that the foreseen approach with intelligent agents could also apply to the problem of bandwidth regulation between the different established connections ('Communications Manager'). Using this approach would allow us to develop the messaging part simply by increasing the defined message types.

The bidirectional regulation of the data traffic is facilitated by the impact of each local agent on the different data transmission threads, which is in turn based on the priority of the different channels, and on the information that is gathered locally and from the agents network.

A fundamental aspect in this development is the definition of reaction mechanisms for asynchronous events, such as changes in the priorities of specific tasks, cancellations, appearance and disappearance of network devices, the triggering of medical alarms (which imply a reconfiguration of priorities that improve the processing and transmission of concrete data related to the patients), and requests of the medical expert.

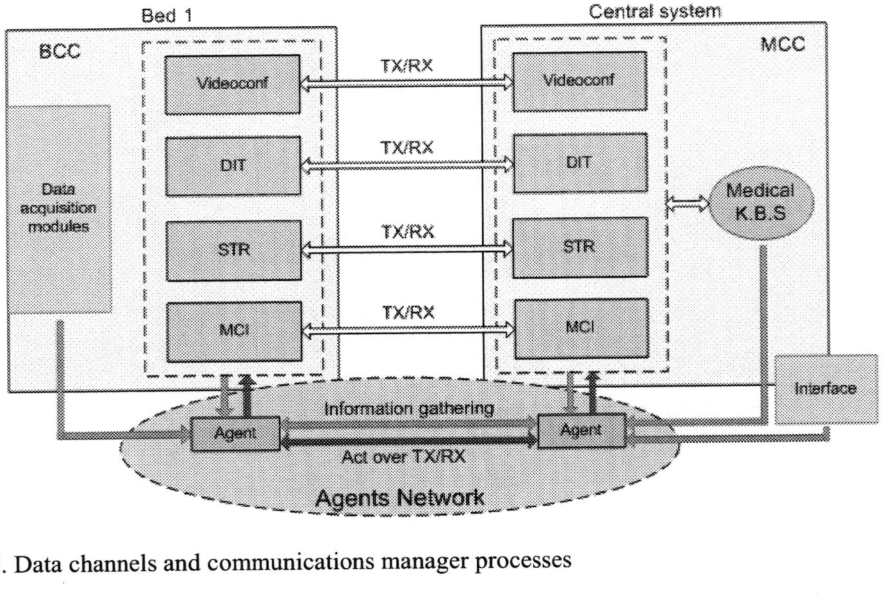

Figure 5. Data channels and communications manager processes

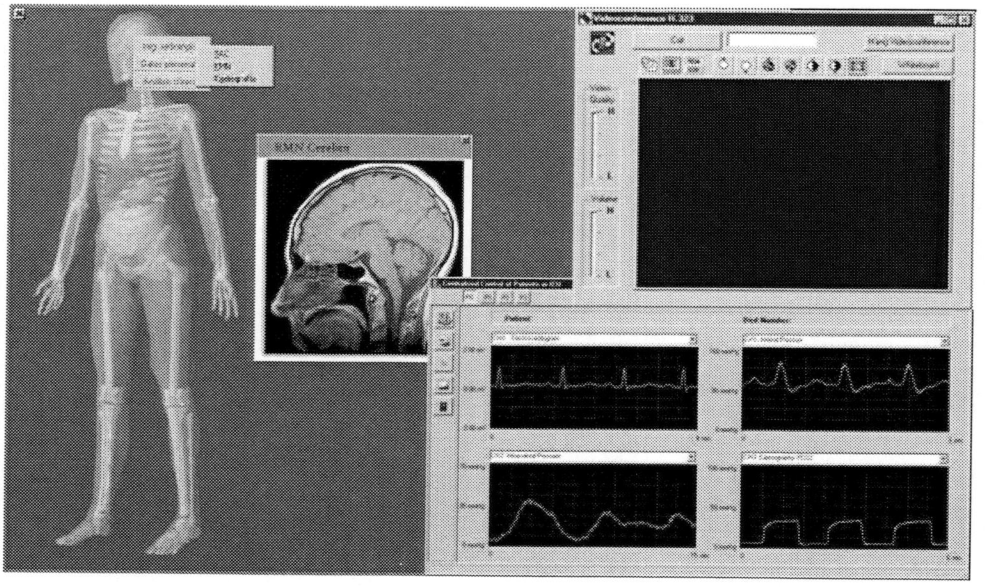

Figure 6. Examples of the monitoring system screens

VIDEOCONFERENCING STANDARDS

Most telemedicine/telehealth applications are equipped with a complete videoconferencing module, which is similar to that of other environments except for the incorporation of data that proceed from peripherals for medical practice: medical imaging devices, patient monitoring devices, etc.

An important factor for these systems is the required image and sound quality, which depends on the concrete application to which the system is assigned. In some cases normal image quality is sufficient (e.g. in teleconsultation, tele-assistance), whereas other cases require high quality images in real time (e.g. telediagnostics, telesurgery); according to the required image and sound quality, users will need specific bandwidth to maintain the quality of their communication.

The videoconferencing systems in the medical/sanitary field have to comply with audio and video standards that guarantee the interoperability of the devices and facilitate their implantation and adaptation to specific tasks. These standards normally use compression formats to adapt to the existing transmission means while maintaining acceptable image and sound quality.

There is a wide range of standards for videoconferencing in medical informatics, but the most commonly used video compression format is usually H.323, based on standard H.261, developed for ISDN communications by the Telecommunication Standardization Sector of the International Telecommunication Union (ITU-T).

Videoconferencing Communication Norms

The ITU-T is the organism that is in charge of coordinating communication standards on a worldwide scale, including the norms that regulate the videoconferencing communications in the medical/sanitary field.

The videoconferencing standards are defined in a series of ITU-T regulations known as "H Series" (audiovisual and multimedia systems) and classified according to the desired quality level into the following categories:

- H.320: Videotelephonic systems and terminals with narrow bandwidth. This norm defines the transport of videoconferencing and vidotelephony over ISDN, offering the appropriate quality for business communications. The narrow bandwidth (N-ISDN) for this specification is defined by ratios that range from 64 kbit/s to 1920 kbit/s using one or multiple ISDN channels. This norm, in turn, uses other standards, particularly standards for the codification of video, audio, and data:
  H.261. Video codec for audiovisual services at a speed of *p* x 64 kbit/s. This norm is used for video transmission.
  G.711, G.722, G.728. Audio codification norms of the "G Series" (Transmission systems and media, digital systems and networks). G.711: Pulse code modulation (PCM) of voice frequencies. G.722: Audio codification of 7 kHz at 64 kbit/s. G.728: Coding of speech at 16 kbit/s using low-delay code excited linear prediction.
  T.120. Data protocol for multimedia conferences. Norm of the "T Series" (terminals for telematics services) that provides a model for the communication of multipoint data in a multimedia communications environment.
- H.321: Adaptation of the H.320 videotelephonic terminals to environments of the digital network of integrated broad bandwidth services (B-ISDN). Videoconferencing communications over ATM (Asynchronous Transfer Mode) with good quality for business-related communications. This version uses the same norms as H.320 for the codification of video, audio, and data.
- H.322: Videotelephonic systems and terminal equipments for local area networks that guarantee the quality of the service. The same norms as H.320 are used for the codification of video, audio, and data.
- H.323: Multimedia communication systems based on packages. Videoconferencing over IP/Ethernet, no guaranteed service quality. Similarly to H.320, this norm uses other standards in turn, particularly the following:
  H.261, H.263, and H.264. Norms used for video transmission. H.263 (video codification) is added to H.320 for communication at low binary speeds. The latest revisions of the H.323 norm have led to the addition of H. 264, a support for the codification and decodification of video (a.k.a advanced video coding for generic audiovisual services).
  G.711, G.722, G.723.1, and G.728. Audio codification norms. With regard to norm H.320, audio norm G.723.1 or a dual rate speech coder for multimedia communications transmitting at 5,3 and 6,3 kbit/s.
  T.120. Data protocol for multimedia conferences
- H.324: Terminal for low bit-rate multimedia communication. Videoconferencing over POTS (*Plain Old Telephone Systems*) with low service quality.

- H.310: Systems and terminals for broad bandwidth audiovisual communications. Videoconferencing over ATM by means of MPEG-2 video compression. This norm provides more quality and is used in applications that require high service quality, as is the case in certain medical applications.

Video Codification Norms

Among the norms that are used for video codification in videoconferencing systems, we would highlight the abovementioned H.261 and H.263 standards. Both are briefly described below.

Standard H.261 can handle *p* x 64 kbit/s (where p=1,2, ..., 30) video sequences, equal to the possible bandwidths in ISDN. It supports two image exploration formats:

- *CIF* (*Common Intermediate Format*). The sampling structure of the luminance consists of 352 image elements per line and 288 lines per image, in an orthogonal disposition. The sampling of each chrominance components counts 176 image elements per line and 144 lines per image, in an orthogonal disposition.

- *QCIF* (*Quarter-CIF*). This format contains a fourth part of the information of a CIF video flow and therefore has half of the image elements and lines of the previous format (luminance: 176 x 144; chrominance: 88 x 72). All the codes must function with QCIF and some can also function with CIF.

The maximum raster ratio for a H.261 video sequence counts approximately 30 frames per second and uses movement compensation to reduce the number of data blocks that are sent in each frame. Also, during codification a video multiplexor reduces even more the flow of transmitted bits.

The H.261 norm is similar to the MPEG standard (*Moving Picture Experts Group*) with regard to DCT codification (*Discrete Cosine Transform*) and quantization, but it changes with regard to the used movement vectors and the presence of future predictions (not applied in H.261).

With respect to communications with narrow bandwidth over POTS, the video standard H.263 was developed as an improvement of standard H.261 with 28,8 kbit/s modem transfers; this norm therefore implies improvements in the movement compensation algorithm and introduces, in a simple manner, forward prediction as a means to obtain a larger compression rate. Three formats are added to image exploration formats H.261, CIF, and QCIF:

- *sub-QCIF:* The luminance sampling structure counts 128 image elements per line and 96 lines per image. The sampling of each chrominance component counts 64 image elements per line and 48 lines per image.

- *4CIF:* This format counts 704 image elements per line and 576 lines per image in luminance, and 352 image elements per line and 288 lines per image in chrominance.

- *16CIF:* This format counts 1408 image elements per line and 1152 lines per image in luminance and 704 image elements per line and 576 lines per image in chrominance.

The video codification norm H.264 was jointly developed by the ITU-T and the ISO/IEC (ISO/IEC 14496-10), and is also known as MPEG-4 AVC (Advanced Video Codec). It was originally created to provide good quality compressed video with bit ratios below those of previous standards (H.323, MPEG-2, and MPEG-4 part 2).

The videoconferencing standard H.323 describes terminals and other entities that provide multimedia communication services over package-based networks (PBN), which in turn may provide a service whose quality is not guaranteed. The H.323 entities can provide audio, real-time video, and/or data communications. The support is obligatory for audio but optional for data and video; however, if it exists, it should have the capacity of using a common operation specification by which all the terminals support these types of means.

The networks used by the H.323 entities to communicate may include local area networks (LAN), wide area networks (WAN), intra-networks, and inter-networks (including the Internet). In turn, these networks can be constituted by simple network segments or various complex topologies that are interconnected by communication links.

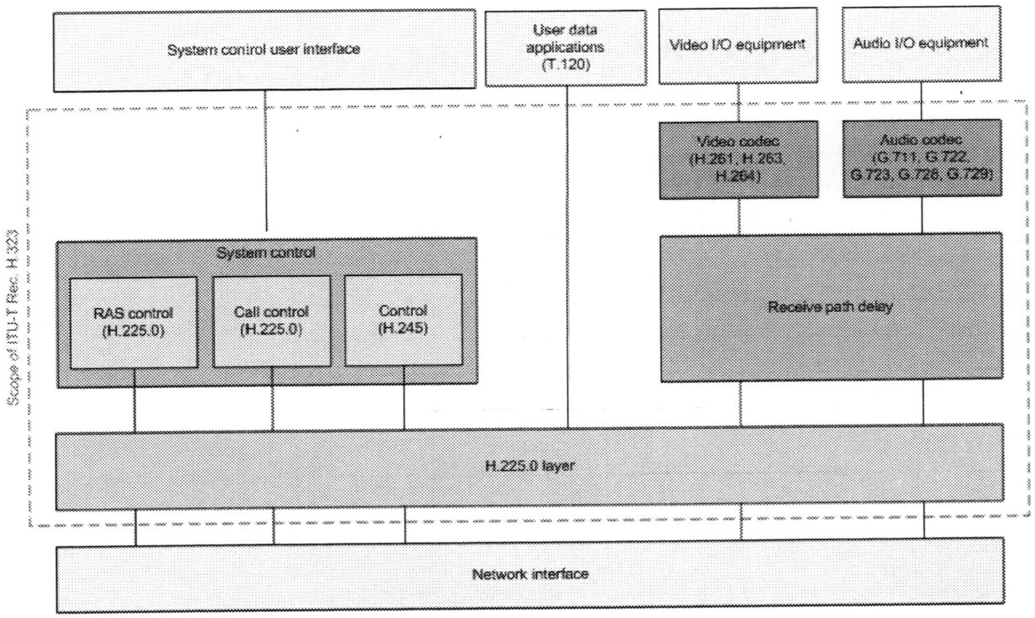

Figure 7. H.323 terminal equipment.

Conclusion

Long-distance medical attention is an old practice that, thanks to new information and communication technologies, has evolved towards the provision of consistent improvements in healthcare. Nowadays, medical data are exchanged in real time through high quality audio and video communications, which contributes to the improvement of the patients' sanitary assistance and quality of life in general. As such, telemedicine is applied in a great many areas, going from teleconsultation, tele-assistance, and telenursing to teleradiology and telesurgery.

Modules for real-time videoconferencing are among the most commonly used methods in telemedicine/telehealth applications, because they follow a series of standards that allow their integration with other devices and facilitate the development of specific tools. Peripheral devices that are connected to computers or videoconferencing equipment increase the interaction and facilitate medical practice.

High costs and poor data quality have for many years obstructed the growth of telemedicine as a full-fledged technology within the healthcare system. In recent decades, however, the fast decrease in costs and technological advances have made telemedicine projects more accessible, even though certain preoccupations with regard to security, confidenciality, efficacy, and guarantees remain valid. The adaptation to the new products is a challenge for healthcare services and telecommunications providers, as well as for the producers of medical equipment and medical information systems.

Future tendencies lean towards the proliferation of more powerful datalines and high definition (HD) audio and video devices that allow the communication between equipments (PCs, mobile devices, mobile videocall-enabled telephones, etc.). The market already offers specific devices for high quality audio and video transmission according to widely acknowledged standards, using audio codecs with up to 320 kbit/s for audio systems 5.1 (AAC - Advanced Audio Coding, formally known as ISO/IEC 13818-7, part of the MPEG-2 and MPEG-4 specification) and high quality video (H.264 full HD, with resolution of up to 1920x1080 and 30fps). The advances in visualization could lead us towards telepresence systems by means of increased virtual reality techniques, providing an appearance of tridimensional presence, improving the visual contact, and raising videoconferencing to an almost real level.

Finally, constant advances in telecommunications, robotics, satellite communications, videoconferencing systems, devices for biomedical data acquisition, medical image devices, etc., in combination with growing social needs, are promising elements for the future of telemedicine. The extension of these types of medical/sanitary applications will make them become part of people's daily lives, because they will dispose of them at home, and the communication between medical staff in remote places will result in a considerable improvement of medical/sanitary attention.

REFERENCES

AIM. (1990). Advanced Informatics in Medicine, Commission of the European Communities. *Supplement Application of Telecommunications of Health Care Telemedicine*, AI, 1685.

American Telemedicine Association. Available from: URL: http://www.atmeda.org

Barnett, G. O., Cimino, J. J., Hupp, J. A. & Hoffer, E. P. (1987). DXplain. An evolving diagnostic decision-support system. *JAMA*, *258(1)*, 67-74.

Bird, K. T. (1975). Telemedicine; concept and practice. *Telemedicine; explorations in the use of telecommunication in health care*. Springfield, Illinois, Thomas.

Burg, G., Soyer, H. P. & Chimenti, S. (2005). Teledermatology. In: Frisch P, Burgdorf W.: EDF White Book, *Skin Diseases in Europe*, 130-133. Berlin, Germany.

CCITT. (1990). *Narrow-band visual telephone systems and terminal equipment. Recommendation H.320*. International Telegraph and Telephone Consultative Committee, *International Telecommunication Union*, Geneva, Switzerland.

CEN. (1993). European Committee for Standardization. *CEN/TC 251* (Health Informatics). Available from: URL: http://www.cen.eu.

Conrath, D. W., Dunn, E. V. & Higgins, C. A. (1983). *Evaluating telecommunications technology in medicine*. Dedham, Massachusetts, USA: Artech House.

Cowain, T. (2001). Cognitive-behavioural therapy via videoconferencing to a rural area. The Australian and New Zealand *Journal of Psychiatry*, *35*, 62-64.

Dafonte, C., Castro, A., Gómez, A. & Arcay, B. (2007). Intelligent agents technology applied to tasks scheduling and communications management in a critical care telemonitoring system, *Computers in Biology and Medicine*, *37(6)*, 760-773.

De Moor, G. (1993). Standardisation in Health Care Informatics and Telematics in Europe: CEN TC 251 Activities. *Progress in Standardisation in Health Care Informatics*, 52-59. Amsterdam, Holland: ISO Press.

Dunn, E. V. & Higgins, C. A. (1984). Telemedicine in Canada; an overview. *Dimensions in Health*, *61(7)*, 16-18.

Eichhorn, I. H., Cooper, I. B., Cullen, D. I., Maier, W. R., Philip, J. H. & Seeman, R. G. (1986). Standards for patient monitoring during anesthesia at Harvard Medical School. *The Journal of the American Medical Association*, *256*, 1017-1020.

ET-ASSIST. (1999). *European Telemedicine for Assistance medicine*. URL: http://www.ehto.org/ht_projects/initial_project_description/et-assist.html

eTEN. (2002). *eTEN European Community programme*. Available from: URL: http://ec.europa.eu/information_society/activities/eten/index_en.htm

EuroPACS. European Society for the promotion of Picture Archiving and Communication Systems in Medicine. Available from: URL: http://www.europacs.org

EuroPaNet Project. (2001). European Psycho-pedagogical Assistance Network. URL: http://europanet.sema.es/

From, S., Stenvold, L. A. & Danielsen, T. (1993). Telemedicine services integrated into a health care network-analysis of communication needs in a regional health care system. *Telektronikk*, *89(1)*, 12-22.

From, S. (1994). Standards for health care telematics - A new dimension and challenge to standards makers. *Telektronikk, special edition Telemedicine*, *1*, 54-63.

Frueh, B. C., Deitsch, S. E. & Santos, A. B., et al. (2000). Procedural and methodological issues in telepsychiatry research and program development. *Psychiatric Services*, *51*, 1522-1527.

Gómez, A., Dafonte, J. C., Arcay, B. & Flores, F. (1999). *A prototype of intelligent telemedicine system in Intesive Care Units*, *17th IASTED International Conference*, Applied Informatics. Innsbruck, Austria.

Guthart, G. S. & Kenneth Salisbury, J. (2000). The Intuitive Telesurgery System: Overview and Application. *Proceedings of the IEEE International Conference on Robotics and Automation*. Detroit. USA.

Health-EU Portal. Available from: URL: http://ec.europa.eu/health-eu/index_es.htm

Hilty, D. M., Yellowlees, P. M. & Nesbitt, T. S. (2006). Evolution of telepsychiatry to rural sites: changes over time in types of referral and in primary care providers' knowledge, skills and satisfaction. *General Hospital Psychiatry, 28(5)*, 367-373.

Hutchison, A., Kaiserswerth, M., Moser, M. & Schade A. (1996). Electronic Data Interchange for Health Care. *IEEE Communications Magazine*, 28-34.

IM. (1996). Institute of Medicine, Committee on Evaluating Clinical Applications of Telemedicine. *Telemedicine: A Guide to Assessing Telecommunications in Health Care.* Washington D.C., USA: National Academy Press.

ISO. International Organization for Standardization. Available from: URL: http://www.iso.org.

Istepanian, R., Laxminarayan, S. & Pattichis, C. S. (2005). M-Health: Emerging Mobile Health Systems. Springer.

ITU-T. (1993). International Telecommunication Union. Recommendation H.261. *Video codec for audiovisual services at p × 64 kbit/s.*

ITU-T. (1996). International Telecommunication Union. Recommendation H.263. *Video coding for low bit rate communication.*

ITU-T. (1998). International Telecommunication Union. Recommendation H.323. *Packet based multimedia communications systems.*

Kampman, J., Hernández, C. & Schwarzer, E. (1989). *Process Control in Intensive Care: An Integrated Working Place*, 1215-1216. Proc. 11th Int. Conf. IEEE EMBS, Seatle, USA.

Katcher, B. S. (2006). MEDLINE: *A Guide to Effective Searching in PubMed and Other Interfaces* (2nd Edition). San Francisco, USA: Ashbury Press.

Kiekens, C., Michiels, K., Machiels, H., Beeckmans, K., Vancoillie, P., Beullens, M. & Lysens, R. (2002). *The effect of treatment by means of the EuroPaNet (European Psycho-pedagogical Assistance Network) on the psychological and medical problems in patients with a stroke or a traumatic brain injury.* 3rd World Congress in Neurological Rehabilitation, Venice, Italy.

Loane, M. A., Corbett, R. & Bloomer, S. E. et al. (1998). Diagnostic accuracy and clinical management by real-time teledermatology. Results from the Northern Ireland arms of the UK Multicenter Teledermatology Trial. *Journal of Telemedicine and Telecare, 4*, 95-100.

Manfredi, L., Shupe, J. & Batki, S. L. (2005). Rural Jail Telepsychiatry: A Pilot Feasibility Study. *Telemedicine and e-Health, 11(5)*, 574-577.

Manguno-Mire, G. M., Thompson, J. W., Shore, J. H., Croy, C. D., Artecona, J. F. & Pickering, J. W. (2007). The Use of Telemedicine to Evaluate Competency to Stand Trial: A Preliminary Randomized Controlled Study. *Journal of the American Academy of Psychiatry and the Law, 35*, 481-489.

Mars, M. & Dlova, N. (2008). Teledermatology by Videoconference : Experience of a Pilot Project. *South African Family Practice, 50(3)*, 70.

Melvin, W. S., Needleman, B. J., Krause, K. R., Schneider, C., Wolf, R. K., Michler, R. E. & Ellison, E. C. (2002). Computer-enhanced robotic telesurgery. Initial experience in foregut surgery. *Surgical Endoscopy, 16(12)*, 1790-1792. New York, USA. Springer.

Perednia, D. A., Gaines, J. A. & Rossum, A. C. (1992). Variability in physician assessment of lesions in cutaneous images and its implications for skin screening and computer-assisted diagnosis. *Archives of Dermatology, 128(3)*, 357-64.

Perednia, D. A. & Brown, N. A. (1995). Teledermatology: one application of telemedicine. *Bulletin of the Medical Library Association, 83(1)*, 42-47.

Pool, S. L., Stonesifter, J. C. & Belasco, N. (1975). *Application of telemedicine systems in future manned space flight.* Lecture held at the 2nd Telemedicine Workshop, Tuscon, Arizona, USA.

Pradhan, M. (1994). *Important Concepts in Telemedicine.* Section on Medical Informatics, Stanford University, Stanford.

Satava, R. M. & Simon, I. B. (1993). Teleoperation, telerobotics, and telepresence in surgery. *Endosc Surg Allied Technol, 1(3)*, 151-153.

Schmid-Grendelmeier, P., Doe, P. & Pakenham-Walsh, N. (2003). Teledermatology in Sub-Saharan Africa. In: Burg G. (ed). Telemedicine and Teledermatology. *Curr Probl Dermatol. Basel, Karger, 32*, 233-246.

Stankonvic, J. A. & Ramamritham, K. (1988). *Hard Real-Time Systems: A Tutorial.* Computer Society Press (IEEE), Washington DC. USA.

Taboada, J. A., Arcay, B. & Arias, J. E. (1998). Real time monitoring and analysis via the medical information bus. Part I. *Medical & Biomedical Enineering. & Computing, 26(5)*, 557-561.

Taboada, J. A., Arcay, B. & Arias, J. E. (1998). Real time monitoring and analysis via the medical information bus. Part II. *Medical & Biological Engineering & Computing, 26(5)*, 557-561.

Thrall, J. H. & Boland, G. (2002). Teleradiology. In: K. J., Dreyer, A. Mehta, & J. H. Thrall, (Eds). PACS: *a guide to the digital revolution.* 315-348. New York, USA. Springer-Verlag.

Thrall, J. H. (2005). Reinventing radiology in the digital age. I. The all-digital department. *Radiology, 236*, 382-385.

Thrall, J. H. (2005). Reinventing radiology in the digital age. II. New directions and new stakeholder value. *Radiology, 237*, 15-18.

TIE. Telemedicine Information Exchange. Available from: URL: http://tie.telemed.org

Timm-Giel Amadou, A., Timm-giel, A., Aust, S., Görg, C., Ehrichs, L., Kus, M. & Wischnewsky, M. B. (2003). UMTS Application Trials: *Teleambulance in the IST Project xMOTION.* IST Mobile Summit. Aveiro, Portugal.

WHO. (1988). World Health Organisation. *Informatics and telematics in health. Present and potential uses.* Geneva.

Wootton, R. & Oakley, A. (2002). *Teledermatology.* Royal Society of Medicine Press Ltd.

XMOTION, *eMobile Testbed For Interoperability Of Net-works In eLogistics.* (2001). IST Project. URL: http://www.ist-world.org

In: Videoconferencing: Technology, Impact and Applications ISBN: 978-1-61668-285-9
Editor: Adam C. Rayler, pp. 81-109

Chapter 4

K-12 EDUCATIONAL VIDEOCONFERENCING APPLICATIONS IN GREECE AND CYPRUS

Panagiotes S. Anastasiades*
University of Crete, Department of Education, 74100 Rethymnon, Greece

ABSTRACT

Information and Communication Technologies (ICT) and especially advanced learning technologies of synchronous transmission can effectively support the educational uses of Video Conferencing features by providing innovative learning and instructional strategies that would improve communication, collaboration and interaction between the learner and the educator. The last 10 years have developed in Greece and Cyprus several important instructional applications via videoconferencing. The objective of this chapter is to present the main educational videoconferencing K-12 applications in Greece.

Keywords: Interactive videoconferencing, Collaboration,

1. INTRODUCTION

A new digital continent is thus born in which concepts like borders, geographical distances, movement or means of transportation, no longer matter. The education sector is the Knowledge Society's main priority, one of the most important portals of the new digital age (Berge & Collins, 1995; Anastasiades, 2002a).

Worldwide, there has been noticed a gap between the knowledge and skills provided by the educational system and the new needs resulting from the Knowledge Society (Adler, 1999; Autor et al, 2000). The various educational systems, through the new curricula and

---

* Correspondign author: Email: panas@edc.uoc.gr, Home Page: http://www.edc.uoc.gr/~panas/

methods, emphasize the collaborative construction of knowledge and the cross-curricular approach so that they link school with authentic real-life situations.

ICT under specific conditions, is expected to provoke numerous, important changes in the contemporary educational environment, as it has already occurred in the wider productive and socio-cultural environment of our times. Communication between students and learners is not dependent on geographical distance; Rather on their ability to access the network (Anastasiades, 2002b; Berge, 2001; Fitzgerald, 2000).

Interactive Videoconferencing is an essential technological medium, which, under pedagogical and social conditions, can significantly contribute to the opening-up of schools to broader social and learning environments. It can also encourage the critical outlook towards the micro world of class, develop collaboration, negotiation, necessity of empathy and the exchange of different attitudes and cultures (Anastasiades, 2009).

The aim of this chapter is to present the main educational videoconferencing K-12 applications in Greece and cyprus.

This chapter is organized as follows:

In the second part we present the educational utilization of IVC on Teaching and Learning. The third part deals mostly with the seven types of IVC in K-12 Instruction and finally in the fourth part we present tha main K-12 educational VC activities in Greece and Cyprus.

2. IVC on Teaching and Learning

Teachers in the 21st century schools are able to utilize advanced learning Internet technologies, digital and school libraries, so that they design environments of collaborative structure of knowledge, emphasizing on cross-curricular approaches, exploratory learning and active participation of students in authentic environments and real-life situations.

The rapid development of ICT applications (synchronous and asynchronous) has established the pedagogical challenges for the World Wide Web ((Lian, 2000, Hammond, 2000)) as one of the most important factors of major and radical changes in the learning environment in which advanced learning technologies meet pedagogy and collaborative learning (Bernard, et al, 2000).

In the recent years, the breakthrough of the Internet, the significant advances in video compression, communication protocols and broadband, along with the increasing need for global communication, have led to the fast development of Videoconferencing service.

IVC technology allows students at two or more distant locations to create a real time collaborative environment (Suthers, 2001; Gibson και Cohen, 2003; Gowan, & Downs,1994; Reed & Woodruff, 1995, Alexander et al., 1999; Chandler and Hanrahan, 2000). Communication may include data and graphics exchange (Brown, 2001; Finn, Sellen, & Wilbur 1997; Andrews, T. & Klease, G. 1998) and data sharing (Gürer, et all 1999).

IVC nowadays has a significant role in order to support collaborative synchronous learning activities at distance (Anastasiades, 2007) (figure 4.6). The Interactive Videoconferencing process allows learners and instructors to share ideas and information, to cooperate, to collaborate at the same time form distant locations (Anastasiades, 2003c; Thompson, 1995; Martin, 2000). This means that emphasis is placed up on the teacher as a cultural intermediary during the learning process in its new form; in the meantime, the social consequences of the new technologies need to be constantly assessed (Anastasiades, 2003d).

IVC under pedagogical and social conditions is an effective tool for the contemporary instructor, as **it** can contribute to the opening up of the class to new communities and the familiarization of students with new learning and cultural experiences and alternative – innovative learning approaches (Anastasiades, 2009).

From a pedagogical perspective, Videoconferencing shapes a new learning and teaching environment which is quite different from the conventional face to face teaching (Hearnshaw, 1999).

Organizing Videoconferencing is rather complicated and this is a serious issue as many factors should be considered: (a) the planning and organizing of the Videoconference, (b) human communication, (c) the specialized education in relation to the continues development of technology and, (d) the high cost of such technological tools (Fincher, 2000).

3. Vc K12 Applications

IVC nowadays plays a significant role in supporting collaborative synchronous distance learning activities (Newman, 2008; Anastasiades, 2007; Greenberg 2006, Gerstein, 2000), promoting interaction in the classroom (Sherry, 1996; Cavanaugh, 2001) as a critical factor to any videoconferencing-based learning situation (Amirian, 2003).

According to Anastasiades (2009) seven types of IVC in K-12 Instruction are frequently found:

1. Experts and Lectures: The objective of this practice is to allow students and educators to be introduced into new ideas and innovations by distinguished experts (scientists, artists, literati, etc.) in order to raise awareness of students on a number of issues (McCombs et al, 2007; Barbanell et al, 2003; Edens, 2001).
2. Virtual Field Trips: offer a class the opportunity to tour a place without commuting (Pachnowski, 2002; Woerner, 1999). A VFT consists a "communication and cultural bridge" for students and schools worldwide regardless of their geographical location (Harkess, 2007; Ashton, 2002; Stainfield et al, 2000).
3. Connecting Schools (Participation Events): The IVC offers two or more schools the opportunity to co-organize events regardless of their geographical location, aiming at the consolidation of relationships and the exchange of cultural and social concepts (King, & Kullman, 2007, Newmman et al, 2005; Cole et al, 2004; Drescher et al, 2005),
4. Instructional Activities at Distance: Through the IVC two or more classes have the opportunity to communicate so that teachers and students share educational techniques, innovative ideas and instructional methodology, etc. (Martin, 2005; Anastasiades, 2003; Gage et al, 2002, Yost, 2001).)
5. Virtual Collaborative Classroom (Cross-curricular thematic approach): The IVC offers two or more schools the opportunity to create, under pedagogical conditions, an environment for collaborative structure of knowledge at distance (Virtual Collaborative Classroom), aiming at the open-up of school towards wider social and

cultural environments (Anastasiades, 2003; 2007; 2009; Bidjerano & Wilkinson, 2008; King, & Macklam, 2007; Ertl et al, 2006; Howland & Wedman, 2003).

6. Distant Locations and Disabilities (Homebound Videoconferencing): This VC type covers mostly students at distant locations or students with disabilities (Belo, Knowlton & Chaffin, 2007; Newman, 2008).
7. Teachers Training & Administration Collaborative activities at Distance: IVC can significantly facilitate teachers' access to training programs and administrative activities regardless of their geographical location (Anastasiades, 2008; Stewart & Vallance, 2007; Kullman & King, 2007; Basham et al, 2006).

Conclusively IVC is a significant tool for DL educational applications, as it contributes to the creation of an interactive environment for communication and collaboration between schools regardless of their geographical location. Nowadays VC tends to become an integral part of the daily routine of numerous schools at all levels.

4. Videoconferencing K-12 Educational Applications in Greece and Cyprus

The first educational VC application in Greece was implemented in 1998 by the 6th Junior High School of Serres, when a teacher took the initiative to connect from home with his students at school. Since then, there have been numerous steps as a result of the rapid development in communication infrastructure, the constant improvement of synchronous transmission technologies and the familiarization of teachers and students with new technologies.

This research into the application of educational VCs was conducted via print or electronic sources on the Internet, proceedings of Greek and International Conventions and journals. In this chapter we present the most important applications of educational VCs in class, according to the classification by Anastasiades (2009), presented in the previous chapter.

4.1. Experts and Lectures

Title:	**Junior High school students discuss with technicians from the Greek Telecommunication Company (1999)**
Content Provider:	6th Junior High-school of Serres
Description:	In October 1999 students from the Web Team of the 6th Junior High-school of Serres participated in a VC with Communication technicians and discussed the new trends in the VC field. The VC was held in the town theatre.
Grades:	6-9
Home Page	(http://www.6gymnasio.gr/ACTIVITIES/THLEMATIKH/thlemat_ote/thlemat_ote.html)

Photos	 Source: http://www.6gymnasio.gr/ACTIVITIES/THLEMATIKH/thlemat_ote/thlemat_ote.html

Title:	**High-school students discuss with Vocational Guidance experts (2000)**
Content Provider:	6th High-school of Serres Municipality of Serres
Description:	On 26th, 27th and 28th June 2000, High-school students and graduates discussed with Vocational Guidance experts. Specialised counsellors, headmasters and cultural attaches from countries with Greek students (USA, G. Britain, France, Spain, Hungary, Canada, Belgium, and Netherlands) answered the students' questions.
Grades:	9-12
Home Page	http://www.6gymnasio.gr/ACTIVITIES/THLEMATIKH/thlemat_wci/thlemat_wci.html
Photos	

Title:	**Update to students on applications of telematics in class (2000)**
Content Provider:	Research and Training Institute of East Aegean
Description:	The Research and Training Institute of East Aegean organized a VC among three schools in Rhodes, Athens and Tirana, to update students on the applications of telematics in class.
Grades:	4-6
Home Page	www.ineag.gr
Photos	-

Title:	**High-school students discuss with the President of the Pedagogical Institute (March 2001)**
Content Provider:	6th Junior High-school of Serres
Description:	In March 2001, the President of the Pedagogical Institute, Pr. Alahiotis, discussed via VC with students from the 6th Junior High-school of Serres about the goals of the P.I. and other current educational issues, such as new subjects, additional instruction etc.
Grades:	

(Continued)

Home Page	http://www.6gymnasio.gr/ACTIVITIES/THLEMATIKH/thlemat_alaxiotis/thlemat_alaxiotis.html
Photos	

Title:	**High-school students participate in a scientific Convention via VC (October 2001)**
Content Provider:	6th High-school of Serres
Description:	Students from the 2nd grade of High-school watched the presentation of a specialized telematics program in the 7th Scientific Convention of Information Technology, held by the Aristotle University of Thessaloniki on October 13th, 2001.
Grades:	9-12
Home Page	http://www.6gymnasio.gr/ACTIVITIES/THLEMATIKH/thelemat_apt_ns/telemat_apt_ns.htm
Photos	

Title:	**Seminar on Earthquake protection (November 2001)**
Content Provider:	Research and Training Institute of East Aegean Ministry of Education – Dpt. Of Elementary Education Greek Communication Company
Description:	On 30-11-2001 students from Trikala, Lakonia and Chania attended a course on earthquake protection, presented by Pr. John Koutsikos, who was based in the VC room of the Greek Communication Company in Athens.
Grades:	-
Home Page	http://users.otenet.gr/~dim2cha/tilekpaideysi.htm
Photos	-

Title:	**An Elementary school in Cyprus communicates via VC with the participants of a scientific convention in Greece (October 2004)**
Content Provider:	University of Crete – Department of Education
Description:	Students from the Elementary school of Agia Fyla in Limassol, Cyprus participated via VC in the scientific convention on DL, held in Rethymno, Crete. The students presented traditional dances and discussed the educational system of Cyprus.
Grades:	5-6
Home Page	http://www.edc.uoc.gr/~panas/LLL/index.htm
Photos	

Title:	**Greek Minister of Education discusses with students about Europe (May 2006)**
Content Provider:	Greek School Network
Description:	On May 9th 2006, the Minister of Education connected via VC with schools around Greece and discussed with the students on the issue "Which Europe do we want to live in?". The Minister answered questions about: The relationship between Europe and Education, the educational opportunities, training, scholarship and student exchange in the E.U. The participating schools were: the 1st Junior High-school of Syros, the 2nd High-school of Edessa, the 5th Junior High-school of Heraklio, the 7th Technical-Vocational School of Larissa, the 15th High-school of Patras, the Experimental High-school of the Aristotle University and the 1st Technical-Vocational School of Argos.
Grades:	7-12
Home Page	http://www.ypepth.gr/docs/9_5_06_d_t1.doc http://vod.sch.gr/video/view/526.html
Photos	

4.2. Virtual Field Trips

Title:	**A Greek school connects with NASA, British Museum**
Content Provider:	Zeridi Schools
Description:	The VC was implemented as an educational tool with the aim of exchanging information and views among educational, scientific and cultural organizations in Europe and the U.S.A, such as NASA, British Museum κλπ.
Grades:	4-6
Home Page	http://www.ziridis.gr/ziridisgrsite/educationprograms/foreignlanguages.html

4.3. Connecting Schools

Title:	**Oikade 2000-2010**
Content provider:	Bank of Cyprus
Description	"Oikade" is an educational programme of the Bank of Cyprus, which develops the possibilities of new technology, so that it brings together Greeks all over the world. Through the Internet and the website www.oikade.gr, it eliminates geographical distance and offers students a Greek website of knowledge and entertainment. Through games, virtual tours and other activities students learn about the Greek culture in an interactive and pleasant way. Moreover, this website hosts a global network of schools, where elementary schools from Greece and Cyprus and Greek-speaking schools around the world develop live net meetings in groups of three. The main theme of these meetings is "Greek neighborhoods around the world". Each class invites the other two to a virtual trip in their region. Students search, collect and present to their peers information and the features that characterize their home region, such as geography, natural resources, history, customs and culture.
Grades	3-6
Home Page:	www.oikade.gr http://www.oikade.gr/contactoik/schools.asp
Photos	

Title:	**«What we think and what the others think» - Exchange of views on Information Technology in Junior High-school (2000)**
Content Provider:	4th Junior High-school of Chania Junior High-school of Platanias, Chania
Description:	The main idea was to connect via net-meeting two classes from different schools, so that they exchange views and discuss joint group or individual projects, already assigned to the students in the previous class session. This activity was awarded the first prize as the "Best educational utilization of new technologies" in the "Internet & Multimedia 2000" competition.
Grades:	7,9
Home Page	http://www.scribd.com/doc/7870760/-
Photos	

Title:	**Students from Greece and Cyprus discuss and exchange views on their educational systems (2001)**
Content Provider:	Technical-Vocational School of Magnisia
Description:	On May 31st 2001 students from Volos and Cyprus discussed via VC, held between the Technical-Vocational School of Magnisia and the Ministries of Education of Greece and Cyprus. It was the first time that secondary

Title:	**Students from Greece and Cyprus discuss and exchange views on their educational systems (2001)**
	education students of Greece participated in a VC. The participating schools were: the 1st Junior High-school of Volos, the 2nd Technical-Vocational School of Ionia, the Cyprian High-school of Nicosia and the Junior High-school of Paralimni. The aim of this activity was to utilize VC technologies and connect students from different social and educational backgrounds, so that they exchange views on educational and other issues of common interest. The organizers' ultimate goal was the establishment of communication via the Internet between schools from Magnisia and Cyprus, aiming at a) the introduction of students and teachers exchange and b) the fraternization of schools in Magnisia and Cyprus.
Grades:	
Home Page	http://1epal-volou.mag.sch.gr/old_site/todays_news/tilediaskeci.htm
Photos	-

Title:	**Students from Serres and Chania discuss the Olympic Games (2003)**
Content Provider:	Greek Communication Company
Description:	In 2003 the Greek Communication Company in Serres organized a VC for schools of the region and Vamos, Chania, where students discussed the Olympic Games.
Grades:	-
Home Page	53 http://www.protiserron.gr/proti2002/DELTIO%20TYPOY.htm
Photos	-

Title:	**Video-conference between Greek and Spanish students (2003)**
Content Provider:	31st Elementary School of Thessaloniki Elementary School of Ceip a Laxe – Marin, Spain
Description:	On April 8th 2003 a VC was implemented between the 31st Elementary School of Thessaloniki (hosted in the VC room of the University of Macedonia) and the Elementary School of Ceip a Laxe – Marin, Spain. The 31st Elementary School of Thessaloniki participates in the Comenius program (www.scuolachiarelli.it/sea), which promotes teachers exchange in 15 E.U participating countries. In this case, the Greek Music teacher visited the Spanish school. The VC was held on demand of the Greek students, who wished to present live the Spanish songs they had been taught by their teacher prior to his departure. The University of Macedonia offered their VC room, which has the necessary audio infrastructure. The Spanish students introduced themselves, thanked the Greek students and promised to present their own choir in Greek songs.
Grades:	5-6
Home Page	http://www.it.uom.gr/projects/cms/index.php?module=content&file=teleconference

(Continued)

Title:	**Video-conference between Greek and Spanish students (2003)**
Photos	

Title:	**SHEDIA (SENA): Video-Conferences between schools on the Aegean Islands (2000-2004)**
Content Provider:	University of the Aegean – Department of Elementary Education Ministry of Education
Description:	The SHEDIA program (School Educational Network of the Aegean) was implemented from 2000 to 2004 on the Aegean Islands (Dodecanese, Cyclades and Eastern Aegean islands). The aim of the program was to upgrade the quality of education via VC and relieve isolation of 46 schools on 36 islands. Example: the students from a multigrade school in Amorgos participated via VC in a Painting competition and also, broadcast their school celebration of March 25th to the school in Psara.
Grades:	1-6
Home Page	http://www.rhodes.aegean.gr/sxedia http://www.ypai.gr/isite/page/1189,1,0.asp?mu=10&cmu=39&thID=0#
Photos	

Title:	**eTwinning (2005-2010)**
Content provider:	European Union's e-Learning programme
Description	eTwinning promotes school collaboration in Europe through the use of Information and Communication Technologies (ICT) by providing support, tools and services to make it easy for schools to form short or long term partnerships in any subject area.
Grades	k-12
Home Page:	www.etwinning.net http://www.etwinning.gr/
Photos	 Πηγή: http://www.etwinning.gr/

Title:	**VC with high-schools from Greece, Cyprus and expatriated students in Australia (2007)**
Content provider:	University of Crete – Department of Education Center of Intercultural and Migration Studies (E.DIA.M.ME.) / program: Greek education abroad
Description	Students from Greece (2nd high-school of Rethymno), Cyprus (High-school of Neofitos – Pafos) and Australia (St. George College – Adelaide) connected via VC (March, 21, 2007. Students presented their home country, home town and their school. They discussed issues of common interest (the educational system, University access, daily life in their countries, etc.) The VC was broadcast live on the internet : http://webcast.ucnet.uoc.gr/?page=View&webcast_id=77
Grades	Grades 9,10
Home Page:	http://www.edc.uoc.gr/~odysseas/
Photos	 VC with high-schools from Greece, Cyprus and Greek origin students in Australia (March 2007)

4.4 Instructional Activities at Distance

Title:	**1998: The 1ST Video-Conference in Greece, 6th Junior High-school in Serres**
Content Provider:	6th Junior High-school of Serres
Description:	The first educational VC application in Greece was implemented in 1998 by the 6th Junior High School of Serres, when a teacher, V. Gipertis, took the initiative to connect from home with his students at school via net-meeting.
Grades:	7-9
Home Page	(http://www.6gymnasio.gr/ACTIVITIES/didask_inter/did_int.htm)
Photos	 Πηγή: http://www.6gymnasio.gr/ACTIVITIES/logos_ellaniforos/logos_ellan.html

(Continued)

Title:	**Instruction for remote schools**
Content Provider:	
Description:	The Research and Training Institute of East Aegean organized a VC among multigrade schools (in Arki, Agathonisi and others) and schools in Athens, aiming at joint instruction.
Grades:	4-6
Home Page	www.ineag.gr
Photos	-

Title:	**ODYSSEAS 2000-2003**
Content provider:	University of Crete, Department of Education University of Cyprus, Department of Computer Science Ministry of Education and Culture of Cyprus Republic
Description:	12 elementary schools of Cyprus participated in the program of educational VCs, called Odysseas 2000-2003. Approximately 300 students and 20 teachers conducted instructions in language, mathematics, physics and environmental studies, based on a drawn pedagogical model (Anastasiades, 2003) - *in the school year 2000-2001,* the pilot phase of the program was implemented in the Elenio elementary school of Nicosia and the elementary school of Agia Fila in Lemessos. The participants were two teachers and 50 students at both schools (Anastasiades, 2003) - *in the school year 2001-2002* , there were VCs in science and mathematics at 4 elementary schools of Cyprus. Science classes: 56 students and 4 teachers from the Elenio school in Nicosia and 19th Koupatio school collaborated to co-organize a Science Celebration (Anastasiades et al, 2003) Mathematics classes: 49 students and 4 teachers from the Elementary school of Agia Fila in Lemessos and the Zinon elementary school in Larnaca conducted distant educational activities via VC in mathematics (Anastasiades et al, 2003) - *in the school year 2002-2003,* instructional activities at distance were implemented in language, mathematics and environmental studies at 6 schools of Cyprus. Language classes: 4 teachers and 37 students from the Elenio elementary school and the school of Agios Antonios in Nicosia conducted distant educational activities via VC in language (Anastasiades, 2006). Environmental studies classes: 4 teachers and 53 students from the elementary schools of Agia Fila and Pissouri in Lemessos conducted distant collaborative activities via VC in environmental studies (Anastasiades, 2006; Chabiaouris, 2008). Mathematics classes: 2 teachers and 51 students from the Zinon elementary school in Larnaca and the 9th elementary school of Pafos conducted distant collaborative activities via VC in mathematics (Anastasiades, 2006)
Grades:	Grades 5,6
Home Page:	http://www.edc.uoc.gr/~odysseas/

Title:	**ODYSSEAS 2000-2003**
Photos	  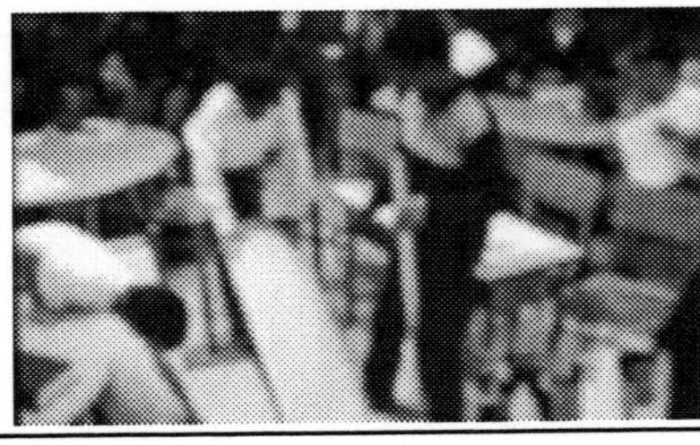

Title:	**Greek language courses for Expatriated students in Varna, Bulgaria (November 2000)**
Content Provider:	World Council of Greeks abroad Information and Telematics Lab of 6th Junior High-school of Serres Unesco Municipality of Serres (Municipal Organization of Cultural and Social Development)
Description :	The aim of the program was the instruction of the basics of the Greek language to expatriated Greek students, and dissemination of Greek culture and customs. The participants were 15-45 students from Varna, Bulgaria.
Grades:	1-6
Home Page	http://www.6gymnasio.gr/ACTIVITIES/logos_ellaniforos/logos_ellan.html
Photos	

Title:	**SHEDIA (SENA): Video-Conferences between schools on the Aegean Islands (2000-2004)**
Content Provider:	University of the Aegean – Department of Elementary Education Ministry of Education
Description:	The SHEDIA program (School Educational Network of the Aegean) was implemented from 2000 to 2004 on the Aegean Islands (Dodecanese, Cyclades and Eastern Aegean islands). The aim of the program was to upgrade the quality of education via VC and relieve isolation of 46 schools on 36 islands. The first VC was implemented among three schools (two in Kremasti, Rhodes and the other in Mandraki, Nisiros). A teacher, based in the Aegean

(Continued)

Title:	**SHEDIA (SENA): Video-Conferences between schools on the Aegean Islands (2000-2004)**
	University facilities in Rhodes taught four lessons in Geography. The software used was Net-meeting and the connection was via ISDN 64K.
Grades:	5
Home Page	http://www.rhodes.aegean.gr/sxedia http://www.ypai.gr/isite/page/1189,1,0.asp?mu=10&cmu=39&thID=0#
Photos	

Title:	**Vocational Guidance Course (2000)**
Content Providers:	Technological Institute of Piraeus Pedagogical Institute (Dpt. Of Vocational Guidance) Secondary Education schools E.A.T.A. (European Association for Telematic Applications)
Description:	The aim of the program is to introduce new technologies into secondary education. The VC was implemented to inform students on vocational guidance issues.
Grades:	7-9
Home Page	http://www.6gymnasio.gr/ACTIVITIES/THLEMATIKH/thlemat_tei_peir/thlemat_tei_peir.html
Photos	  The 6th Junior High-school of Serres  The Videoconferencing Room of TEI Piraeus

4.5. Virtual Collaborative Classroom

Title:	**ODYSSEAS 2004: Olympic games- cross curricula approach**
Content provider:	University of Crete, Department of Education University of Cyprus, Department of Computer Science Ministry of Education and Culture of Cyprus Republic
Description:	The European Parliament and European council decided, after the recommendation of the European committee, the designation of the year

Title:	**ODYSSEAS 2004: Olympic games- cross curricula approach**
	2004 as "the European year of Education through sports", in an attempt to emphasize on the educational and social value of sports and their role in combating racism and xenophobia. This action in combination with the historical return of the Olympic Games to their birthplace led us to the decision to focus attention on the cross-curricular approach of Olympic Education, taking advantage of the DL methodology of the Odysseas program (Anastasiades, 2004; 2006) The participants in "Odysseas 2004 – Olympic Education" were ten schools: 8 in Cyprus and 2 in Greece (25 teachers and approximately 300 students). Each pair of schools was assigned one of the following modules: Module 1: Ancient Olympic Games Module 2: Modern Olympic Games Module 3: Olympic Games – Athens 2004 Module 4: Paralympics Module 5: Despina and the dove In chapters 10-14 we present all modules in detail. The pedagogical model and evaluation methodology of Odysseas were considered as one of the best programs – practices at European level and were publicized in the database "Best Europe Practices knowledge based database" (http://www.beep-eu.org/).
Grades:	5,6
Home Page:	http://www.edc.uoc.gr/~odysseas/
Photos	

Title:	**ODYSSEAS 2006: The sea unites us**
Content Provider:	University of Crete, Department of Education
Description:	The focal point of the ODYSSEAS 2005-2006 program is the development and implementation of DL collaborative learning activities via VCs between elementary schools. The aim of the program is to raise awareness, inspire and motivate the elementary participating students to adopt attitudes and values, promote

(Continued)

Title:	**ODYSSEAS 2006: The sea unites us**
	collaboration, understanding, autonomy, critical thinking and responsibility according to the principles of pluralism. The theme of the program was "The sea unites us". - In 2005-2006 8 classes from 7 Elementary schools participated in the program: 4 schools in Crete (Heraklio, Rethymno, and Chania), one in the island of Aegina and two in Athens. - The participants were 21 teachers and 220 students. - The modules/topics were: Module A: "Sea experiences" (1st Elementary school of Aegina and 2nd Elementary school of Rethymno) Module B: "Sea and History" (1st Elementary school of Arhanes, Heraklio and 1st Elementary school of Aegina) Module C: "Sea and History" (Elementary school of the naval yard in Souda, Chania and 1st Experimental Elementary school of Athens) Module D: "Sea voyages" (2nd Experimental Elementary school of the University of Athens and 6th Elementary school of Rethymno)
Grades:	5,6
Home Page	http://www.edc.uoc.gr/~odysseas/
Photos	

Title: Program Flyer	**ODYSSEAS 2007**
Content Provider:	University of Crete, Department of Education
Description:	The focal point of the ODYSSEAS 2007 program is the development and implementation of cross-curriculum DL collaborative learning activities via VCs between elementary schools. - In 2006-2007 8 teachers and 120 students participated in the program: two schools were in Crete (Rethymno, Chania), one in the island of Aegina and one in Athens. - The ODYSSEAS 2007 program was divided into two modules:

	A. The first module focuses on the Mediterranean Sea and aims at raising awareness on environmental issues through DL material specially designed for elementary school students. The participants were the Elementary school of the naval yard in Souda, Chania and the Leontio Elementary school of Athens. B. The second module focuses on local contemporary biographies and aims at familiarizing the students with different local cultures through cross-curriculum approach and by promoting collaboration skills and critical thinking. The participants were the 1st Elementary school of Aegina and the 2nd Elementary school of Rethymno.
Grades:	5,6
Home Page	http://www.edc.uoc.gr/~odysseas/
Photos	

Title: Program Flyer	**ODYSSEAS 2008**
Content Provider:	University of Crete, Department of Education
Description:	In this program emphasis is laid upon collaborative learning and the development of social and technological skills, so as to prepare the future citizens for their creative integration into the emerging Society of Knowledge of the 21st century. The participants were 9 teachers and 120 students: two schools were in Crete (Chania, Kissamos), one in the island of Aegina and one in Athens. The ODYSSEAS 2008 program was divided into two modules: A. The first module focuses on the Mediterranean Sea and aims at raising awareness on environmental issues through DL material specially designed for elementary school students. The participants were the 1st Elementary school of Aegina and the 3rd Elementary school of Kissamos.
	B. The second module focuses on climatic change. According to the suggested methodology and by implementing the animation and script-writing techniques, the students participated in real-life learning situations, so as to develop interest in environmental issues and take initiatives for the protection of the environment. The participants were

(Continued)

	the Elementary school of the naval yard in Souda, Chania and the Leontio Elementary school of Athens.
Grades:	5,6
Home Page	http://www.edc.uoc.gr/~odysseas/
Photos	

4.6 Distant Locations and Disabilities

Title:	DAEDALOS: VC environment for the deaf
Content Provider:	Multimedia and Net applications Laboratory of DEMOKRITOS.
Description:	The environment is based on advanced VC web services (network virtual classroom) and offers numerous applications and tools, which are user-friendly and support training in DL instruction for the deaf. The learning material includes text and multimedia for instruction of the English language; it is fully transcribed into sign language digital video and is available on demand of the user (Video on Demand).
Grades:	-
Home Page	http://imm.demokritos.gr/dedalos

Photos	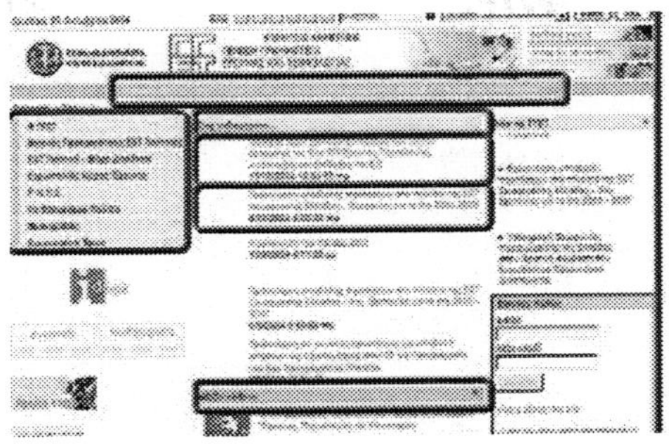

Title:	**VC between Elementary school students and hospital-bound children (October 2002)**
Content Provider:	Bank of Cyprus – OIKADE
Description:	On October 11th 2002 a VC was implemented between the students of 6th grade of the Elenio Elementary school and in-patient children in the Children's Oncology Clinic of Makario Hospital. The in-patient children had the opportunity to discuss with the students, who sang for them and performed Karagiozis (a Greek buffoon type).
Grades:	5
Home Page	http://www.lygeros.org/Interviews/H_Shmerinh_20021012.html
Photos	

4.7. Teachers Training & Administration Collaborative Activities

Title:	**ODYSSEAS 2003: Teachers from Greece discuss with the Library Director of the University of Cyprus**
Content Provider:	University of Crete – Department of Education
Description:	On December 16th 45 teachers of Elementary Education from the University of Crete in Rethymno connected via VC with the University of Cyprus in Nicosia and discussed with the Library Director on the issue: "Digital Libraries in the Information Society: is this the end of the book era?"
Grades:	-
Home Page	**http://www.edc.uoc.gr/~odysseas/**

(Continued)

Photos	

Title:	**DIAS: Teachers' training in multigrade and remote Greek schools (2003)**
Content provider:	Greek-German School School of Humanitarian Studies of the Aegean University
Description:	DIAS project (Satellite Network of Remote Schools) started in December 2003, six months after the launch of the first Greek satellite (HellasSat), and was completed in 2005. it was funded by the Department of Research and Technology, in the context of the program "Electronic Learning" (Apostolakis et al, 2006; 2009). For the project, advanced technological solutions were developed for the inter-school training of Elementary teachers stationed at multigrade schools (Little, 2006), but also for the support of everyday instruction at these schools. During the project, there was a satellite link between the transmission point at the Czommunication Company premises in Marousi, Athens and 13 reception points around Greece: 10 remote schools (in Karpathos, Amorgos, Rhodes, Kefalonia, Crete, Messinia, Epirus, Evros, N. and S. Pindos), one in INTRACOM for the on-line technical support, one in the Greek-German School and one in the Aegean University, where the Research team recorded the development of the course and supported the procedure. The development of the training project was based on the approach suggested by MUSE (Tsolakidis et al, 2003), as well as the Alternative Professional Development Model (Makrakis, 1996; 2004). At the same time, the adopted approach was trainee-friendly (Tsolakidis et al, 2005).
Home Page:	http://www.dias.ea.gr/
Photos	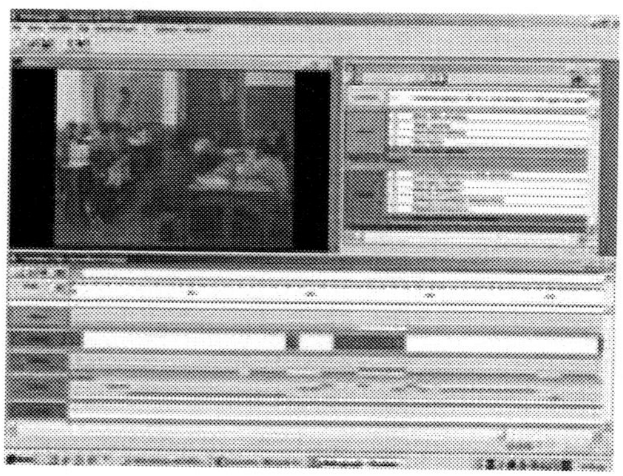

Title:	**ODYSSEAS 2006: Teachers from Greece discuss with a Special Education teacher from Cyprus**
Content Provider:	University of Crete – Department of Education
Description:	On March 8th 2006 30 Elementary teachers from the University of Crete (Rethymno) connected via VC with the University of Cyprus (Nicosia) and discussed with the Special Education teacher, Ms. Eleni Ftiaka, on the contemporary trends of Special Education in Europe.
Grades:	-
Home Page	http://www.edc.uoc.gr/~odysseas/
Photos	 

Title:	**«PAIDEIA OMOGENWN» (EDUCATION OF GREEKS TEACHERS IN DIASPORA (2005-2008)**
Content provider:	University of Crete - Department of Education Center of Intercultural and Migration Studies (E.DIA.M.ME.)
Description:	The whole project «PAIDEIA OMOGENWN» aims to continue, develop and promote the Greek language and culture, among primary and secondary students and teachers of Greek origin, who live and study abroad, as well as non-Greek speaking students who want to learn the Greek language and wish to encultuvate in Greek culture. (Damanakis, 1987). One of the most important topics of the project concerns the implementation of a complete e-learning environment which is adopted and developed in order to provide continuous training to the teachers who teach Greek as a second and foreign language through e-learning. The proposed methodology (Anastasiades, 2005; 2008) aims to create a blended learning environment, which embraced the functional combination (Blended Instructional Model) of educational technologies of synchronous (videoconferences) and asynchronous learning (web based learning platform) [figure 1] in order to provide an interactive learning environment The innovative aspect of the suggested method is threefold (Anastasiades, 2005; 2008): 1. The development of a holistic pedagogical model that supports the functional combination of synchronous and asynchronous learning. 2. The creation of an asynchronous platform that is based on freeware software. 3. The design and improvement of videoconferencing pedagogy according to particular needs of the instructing method that was developed. The proposed model is the cornerstone of the designing and application of a significant DE program for the training of teachers throughout the world, which is implemented by EDIAMME of the University of Crete and is under constant assessment and adaptation. Since 2005 five hundred teachers participated to the training project from - Australia (Melburne, Sidney, Adelaide & Renmark) - Europe (Germany, Russia, Sweden, Denmark, Turkey, Egypt, Belgium, Czech Republic, Georgia, Great Britain & Holland)

(Continued)

Title:	**«PAIDEIA OMOGENWN» (EDUCATION OF GREEKS TEACHERS IN DIASPORA (2005-2008)**
	- United States of America(Chicago, New Port Richey FL , Dallas, Brunswick, Wilmington, Montreal, Mundelein, New York, South Yarmouth, Philadelphia, Brooklyn, Levittown, Boston, East Norwich)
Home Page:	http://elearn.edc.uoc.gr http://www.ediamme.edc.uoc.gr/diaspora/
Photos	**Australia**   **Snapshot**: VC Between Crete (Greece) & Melbourne (Australia) Instructor: M. Damanakis Course: Intercultural Pedagogy December 2005 **Snapshot**: VC Between Crete (Greece) and Melbourne, Sydney &Adelaida (Australia) Instructor: G.Katsimali Course Language March 2006 **Europe**   **March 2006:**Prof Xatzidaki Audience in: Bilefeld & Düsseldorf (Germany) Course: Language **Picture : December 2006:** Prof Katsimali (Course: Language) Audience in: Bilefeld & Düsseldorf (Germany) Live Webcast Audience in 7 countries: Germany, Sweden, Turkey, Netherlands, Belgium, Czech Republic, Georgia , Egypt, United Kingdom, Russian Federation, Luxembourg, Denmark **USA- Canada**   April 2008 Prof Damanakis and Prof Xatzipanagiotidi Audience in New York, Chicago, Washigton, Florida, New Port Richey, Montreal

Title:	**Teachers' training and Learning Communities (2008)**
Content provider:	Greek School Network
Description:	In the context of Modern Distant Learning, the Greek School Network offers Greek teachers the opportunity to implement multi-point VCs via the Click-to-meet software. The participants were 121 Elementary and Secondary Education teachers. The instructional approach of the training focused on the VC as a tool for collaborative learning and creativity (Labropoulos et al, 2009). The modules referred to the usage of Web tools which could be employed within the Project Method context in Elementary and Secondary Education. They were: a) Introduction to the Project Method b) Blog c) Wiki d) Video-Conference.
Home Page:	http://rts.ach.sch.gr/clicktomeet
Photos	 *Snapshot: VC between Instuctors*

Title:	**ODYSSEAS 2008: Teachers' training in Greece and Cyprus**
Content provider:	University of Crete - Department of Education
Description:	The training program was organized by the Dpt. Of Elementary Education of the University of Crete in March-April 2008. There were 3 VCs, in which 65 teachers participated from Athens and Rethymno, Greece and Pafos, Cyprus. On March 27th 2008, Pr. Anastasiades P. from Rethymno presented the topic: "Advanced Learning Technologies in Life-Long and Distant Learning". On April 3rd 2008, Dr. Labropoulou N. from London presented the topic: "Collaborative Learning and the Internet". N April 10th 2008, Pr. Spantidakis J. from Rethymno delivered a lecture on: "Educational Multimedia in Instruction and Learning".
Home Page:	

(Continued)

Title:	**ODYSSEAS 2008: Teachers' training in Greece and Cyprus**
Photos	

5. CONCLUSIONS

The educational utilization of VC has been marking significant progress in the Greek and Cyprian educational systems in the last years. Since 1998, when the first VC was implemented in education, numerous applications have been designed and implemented in K-12 Classrooms in Greece and Cyprus.

Based on the research conducted for this paper, we presented 37 occasions of pedagogical utilization of VC in Greek and Cyprian schools. According to the classification presented, 8 out of the 37 applications were connections with experts (VC experts), one was a Virtual field trip, 8 were connections between schools or participation events, 6 were pertinent to Instructional Activities, 5 were Virtual Collaboration Classrooms, 2 were for people with disabilities and, finally, 7 concerned teachers' training.

From the above classification, we can conclude that there is insufficiency of Virtual field trips, which can be justified by the fact that museums, libraries and other organizations in Greece and Cyprus have not yet developed the necessary infrastructure and content so as to utilize VC as a communication tool with people regardless of their geographical location. In addition, there is room for progress in the field of people with disabilities, as the VC is an effective means for their equal integration into society and the job market.

The familiarization of teachers and students with ICT, the advances of VC systems in connection with the development of broadband have significantly promoted the pedagogical utilization of VC in Greek and Cyprian schools since 2003-2004. However, in most cases, the pedagogical frame of the utilization of VC in class has not been defined yet, which should the subject of our concern in the following years.

REFERENCES

Adler, Richard, P. (1999). *Information Literacy: Advancing Opportunities for Learning in the Digital Age*. The Aspen Institute Forum on Communications and Society, Washington, D.C.

Alexander W, Higgison C., &. Mogey, N (Eds.). (1999). *Videoconferencing in teaching and learning: Case studies*. LTDI and TALiSMAN, Institute of Computer-Based Learning, Heriot-Watt University Edinburgh, SCT. Retrieved June 4, 2004 from www.icbl.hw.ac.uk/ltdi.

Amirian, S. (2003). Pedagogy and Video conferencing: A Review of Recent Literature, In: Greenberg, A. *Navigating the Sea of Research on Video conferencing-Based Distance Education*. Wainhouse Research, Polycom, Inc.

Anastasiades P. (2006a). Interactive Videoconferencing in Lifelong Learning: Methodology and Implementation Policy at the University of Crete (E.DIA.M.ME). In *Proceedings of the Diverse 2006, 6th International Conference on Video and Videoconferencing in Education*. Scotland : Glasgow Caledonian University, Glasgow 5-7 July 2006.

Anastasiades, P. (2002a). The Theory of Information Reversal. *Computers and Society, ACM Special Interest Group on Computers and Society (ACM SIGCAS)*, *32(2)*,10-16.

Anastasiades, P. (2002b). Towards the Global Information Society: The Enactment of a Regulatory Framework as a Factor of Transparency and Social Cohesion. *Lecture Notes in Computer Science, Springer-Verlag Berlin Heidelberg, 2510*, 527-535.

Anastasiades, P. (2003c). Building up a pedagogical model for synchronous distance learning courses. In J. A. Chambers, (Ed.), *Selected Papers Publication from the 14th International Conference on College Teaching and Learning*, April 2003. Jacksonville, FL: Florida Community College at Jacksonville.

Anastasiades, P. (2007). Interactive Videoconferencing (IVC) as a Crucial Factor in Distance Education: Towards a Constructivism IVC Pedagogy Model under a cross curricular thematic approach. In E. Bailey, (Ed) *Focus on Distance Education Developments.*, NY: Nova Science Publishers, Inc

Anastasiades, P. (2007). Interactive VideoConferencing (IVC) as a Crucial Factor in Distance Education: Towards a Constructivism IVC Pedagogy Model under a cross curricular thematic approach. In E. Bailey, (Ed), *Focus on Distance Education Developments*. NY: Nova Science Publishers, Inc

Anastasiades, P. (2008). Blending Interactive Videoconferencing and Asynchronous Learning in Adult Education: Towards a Constructivism Pedagogical Approach. A Case Study at the University of Crete (EDIAMME). In S., Negash, M., Whitman, A. Woszczynski, & K. Hoganson, (Editors). *Handbook of Distance Learning for Real-Time and Asynchronous Information Technology Education*, NY, IGI Global

Anastasiades, P. (2009). *Interactive Videoconferencing and Collaborative Distance Learning for K-12 Students and Teachers: Theory and Practice* NY: Nova Science Publishers, Inc

Apostolakis, E., Koulouris, P., Sotiriou, S. & Tsolakidis, K. (2008). The DIAS program: Teachers' Training in multigrade schools via VC and Satellite technology. In: P. Anastasiades, (Ed.) *The VC in Life-long and Distant Learning: Pedagogical Applications of Collaborative DL*, Athens: Guttenberg Editions.

Apostolakis, et al (2006). Bridging the Digital Divide: The Rural Wings Project. *EDEN Annual Conference 'E-Competences For Life, Employment And Innovation: "E" is more! - E-learning Enabling Education in Evolving Europe.* Vienna: EDEN.

Ashton, T. (2002) New virtual field trips. *Roeper Review, 24(4)*, 236-238

Autor, D., Levy, F. & Murane, R. (2000). *The Skill Content of Recent Technological Change: An Empirical Exploration*. National Bureau of Economic Research, Washington, D.C.

Barbanell, P., Falco, J. & Newman, D. (2003). New Vision, New Realities: Methodology and Mission in Developing Interactive Videoconferencing Programming. Retrieved http://www.archimuse.com/mw2003/papers/barbanell/barbanell.html.

Basham, J., Lowrey, A., Jones, M. & Huffman, D. (2006). Making Use of the Net: Internet Based Videoconferencing and Online Conferencing Tools in Teacher Preparation. In C. Crawford, et al. (Eds.), *Proceedings of Society for Information Technology and Teacher Education International Conference 2006*, (1440-1444). Chesapeake, VA: AACE

Bello, A. D., Knowlton, E. & Chaffin, J. (2007). Interactive Videoconferencing as a Medium for Special Education: Knowledge Acquisition in Preservice Teacher Education. *Intervention in school and clinic*, *43(1)*, 38-46.

Berge, Z. & Collins, M. (1995a). Computer mediated Communication and the Online Classroom in Higher education. *Computer –Mediated Communication Magazine*, vol.2, no.2

Berge, Z. L. & Mrozowski, S. (2001). Review of Research in Distance Education 1990 to1999.*American Journal of Distance Education, 15(3).*

Bernard, R., Rubalcava, B. & Pierre, D. (2000). Collaborative online distance learning: issues for future practice and research. *Distance Education, 21(2)*, 260-277.

Bidjerano, T. & Wilkinson, D. (2008). Designing and Implementing Collaborative Clasroom Videoconferences. In D., Newman, J., Falco, S. Silverman, & P. Barbanell, (Eds). *Videoconferencing Technology in K-12 Instruction, Best Practices and Trends.* Information Science Reference, Hersley- New York

Brown, J. S., Collins, A. & Duguid, P. (1989). Situated cognition and the culture of learning, *Educational Researcher*, *18(1)*, 32-42. Full Text via CrossRef.

Cavanaugh, C. S. (2001). The Effectiveness of Interactive Distance Education Technologies in K-12 Learning. *International Journal of Educational Telecommunications, 7(1)*, 73-88.

Chandler, G. & Hanrahan, P. (2000). Teaching using interactive video: creating connections. *Journal of Nursing Education*, *39*, 73-80.

Cole, C., Ray, K. & Zanetis, J. (2004). *Videoconferencing for K–12 classrooms: A program development guide.* Washington: ISTE.

Edens, K. M. (2001). Bringing Authentic K-12 Classrooms and Teachers to a University Classroom through video conferencing. *Journal of Computing in Teacher Education, 17(3).*

Ertl, B., Fischer, F. & Mandl, H. (2006). Conceptual and socio-cognitive support for collaborative learning in videoconferencing environments. *Computers & Education.Volume*, *47*, Issue 3, 298-315.

Fincher, S. & Petre, M. (1998). *Project-Based Learning Practices in Computer Science Education, FiE Conference*, 1185-1191. IEEE, November 1998.

K. E., Finn, A. J. Sellen, & S. B. Wilbur, (Eds.). (1997). *Video-mediated communication.* Mahwah, NJ: Lawrence Erlbaum Associates.

K. E., Finn, A. J. Sellen, & S. B. Wilbur, (Eds.). (1997). *Video-mediated communication.* Mahwah, NJ: Lawrence Erlbaum Associates.

Fitzgerald, M. (2000). *Educational Media & Technology Yearbook 2000,* Volume 25. Englewood, CO:Libraries Unlimited.

Gage, J., Nickson, M. & Beardon, T. (2002, September) *Can videoconferencing contribute to learning and teaching? The experience of the Motivate project.* Paper presented at the Annual Conference of the British Educational Research Association, University of Exeter, England.

Gerstein, R. (2000). Videoconferencing in the Classroom: Special Projects toward Cultural Understanding. Computers in the Schools, vol. 16, issue 3/4, 177-186.

C. Gibson, & S. Cohen, (Eds.) (2003) *Virtual Teams That Work: Creating Conditions for Effective Virtual Teams.* San Francisco. CA: Jossey-Bass/Wiley.

Gowan, J. A. & Downs, J. M. (1994). Video conferencing human-field study. *Information and Management, 27,* 341-356.

Greenberg, A. (2006). Taking the Wraps off Videoconferencing in the U.S. Classroom A. *State-by-State Analysis Wainhouse Research.*

Gürer, D., Kozma, R. & Millán, E. (1999). Impact of shared applications and implications for the design of collaborative learning environments. In S. Lajoie, (Ed.), *Proceedings for the ninth world conference on artificial intelligence in education* (AI-ED99, 439-445). Amsterdam: IOS Press.

Hambiaouris K., Raptis A., Rapti A. & Anasatsiades P. (2009). A complex instruction in the context of a hybrid-collaborative learning environment applying Web technologies, *Open Education, Vol. 5, issue 1,* 88-100.

Harkess, R. L., Kuehny, J. S., Evans, M. R., Greer, L. & Cavins, T. (2007). Virtual field trips: expanding the greenhouse classroom. Acta Hort. (ISHS) 762:439-444. Retrieved March 14, 2008, from http://www.actahort.org/books/762/762_45.htm

Hearnshaw, D. (1998). Capitalising on the Strengths and Availability of Desktop Videoconferencing. *Active Learning, vol. 7,* 52-59.

Howland, J. & Wedman, J. (2003). Experiencing diversity: Learning through videoconference technology. In *Proceedings of ED-MEDIA 2003 World Conference on Educational Multimedia, Hypermedia & Telecommunications(1),* (1562-1565).

King, C. & Kullman, C. (2007). Integrating Videoconferencing into K-12 Teaching and Learning Using the SuperNet. In C. Crawford, et al. (Eds.), *Proceedings of Society for Information Technology and Teacher Education International Conference 2007,* (1561-1568). Chesapeake, VA: AACE.

King, C. & Macklam, C. (2007). Constructivist and Collaborative Perspectives on Building A Provincial Network in Support of Teaching and Learning Using Videoconferencing in Alberta. In C. Montgomerie & J. Seale, (Eds.), *Proceedings of World Conference on Educational Multimedia, Hypermedia and Telecommunications 2007,* (2059-2066). Chesapeake, VA: AACE.

Kullman, C. & King, C. (2007). A Professional Development Model to Integrate Videoconferencing in K-12 Education. In C. Crawford, et al. (Eds.), *Proceedings of Society for Information Technology and Teacher Education International Conference 2007,* (3073-3076). Chesapeake, VA: AACE.

Labropoulou, N., Kampilis, P., Minaoglou, N., Papadimitriou, S., Vivitsou, M., Konetas, D., Ikonomakos, I., Gikas, A., Paraskevas, M. & Grigoropoulos, E. (2009). It is not what the tools can do for us but what we can do with the tools: VC systems and their Pedagogical Utilization in e-Learning. In: P. Anastasiades, (Ed.) *The VC in Life-long and Distant Learning: Pedagogical Applications of Collaborative DL*, Athens: Guttenberg Editions.

Lian, A. (2000). Knowledge transfer and technology in education: toward a complete learning environment. Retrieved from http://ifets.ieee.org/periodical/vol_3_2000/v_3_2000.html

Little, A. W. (2006). *Education for All and Multigrade Teaching: Challenges and Opportunities*. Dordrecht: Springer

Makrakis V. (2004). Training Teachers for a Sustainable Future Supported by Information and Communication Technologies. In: P. Anastasiades, (Ed.) *1st Pan-Hellenic Convention: Life-long and Distant Learning in the Information Society*, (143-151). Rethymno: Propompos Editions.

Martin, M. (2000). *Videoconferencing in teaching and learning—case studies and guidelines*. Western Education and Library Board.

Martin, M. (2008). Interactive Videoconferencing into the Classroom: A Prespectice from Northern Ireland. In D., Newman, J., Falco, S. Silverman, & P. Barbanell, (Eds). *Videoconferencing Technology in K-12 Instruction, Best Practices and Trends*. Information Science Reference, Hersley- New York.

McCombs, G. B., Ufnar, J. A. & Shepherd, V. L. (2007). The Virtual Scientist: Connecting University Scientists to the K-12 Classroom through Videoconferencing. *Advances in Physiology Education. 31(1)*, 62-66.

Newman, D., Barbanell, P. & Falco, J. (2005). Documenting Value Added Learning Through Videoconferencing: K-12 Classrooms' Interactions with Museums. In G. Richards, (Ed.), *Proceedings of World Conference on E-Learning in Corporate, Government, Healthcare, and Higher Education 2005*, (389-401). Chesapeake, VA: AACE.

Newman, S. (2008). Videoconferencing and the K12 Classroom: What is it? And Why do it? In D., Newman, J., Falco, S. Silverman, & P. Barbanell, (Eds). *Videoconferencing Technology in K-12 Instruction, Best Practices and Trends*. Information Science Reference, Hersley- New York.

Newman, S. (2008). Videoconferencing and the K12 Classroom: What is it? And Why do it? In D., Newman, J., Falco, S. Silverman, & P. Barbanell, (Eds). *Videoconferencing Technology in K-12 Instruction, Best Practices and Trends*. Information Science Reference, Hersley- New York.

Pachnowski, L. M. (2002). Virtual field trips through video conferencing. *Learning and Leading with Technology, 29(6)*, 10-13.

Reed, J. & Woodruff, M. (1995). An introduction to using videoconferencing technology for teaching. The Distance Educator Newsletter, Fall. Retrieved from: http://www.kn.pacbell.com/wired/vidconf/Using.html.

Stainfield, J., Fisher, P., Ford, B. & Solem, M. (2000). International virtual field trips: a new direction?. *Journal of Geography in Higher Education, 24(2)*, 255-262.

Stewart, M. & Vallance, M. (2007). The impact of synchronous inter-networked teacher training in Information and Communication Technology integration. *Computers & Education In Press, Corrected Proof.*

Suthers, D. (2001). Collaborative representations: Supporting face to face and online knowledge-building discourse. *Proceedings of the 34th Hawai`i International Conference on the System Sciences* (HICSS-34), January 3-6, 2001, Maui, Hawai`i: Institute of Electrical and Electronics Engineers.

Thompson, R., Rowan, L. & Kindt, I. (1995). Innovations in videoconferencing: embracing technology and joining regions in university teaching. *CQ Journal of Regional Development, 3(4)*, 57-60.

Tsolakidis, C., Fokides, E., Fokiali, P., Saridou, X., Sotiriou, S., Orfanakis, M., Savvas, S., Little, A., Pridmore, P. & Greig, L. (2003). The MUSE project: A Pilot Programme of Multi-grade Schoolteachers' In-Sedrvice Training. In A. Szuks, & E. Wagner, (Eds.), *The Quality Dialogue. Integrating Quality Cultures in Flexible, Distance and eLearning, EDEN annual conference*, (334-341). Rhodes: EDEN.

C., Tsolakidis, S. Sotiriou, & P. Koulouris, (Eds) (2005). Lifelong e-Learning for Multigrade School Teachers. *Proceedings of the First NEMED Workshop*, Helsinki. Athens: EPINOIA.

Woerner, J. J. (1999). Virtual fieldtrips in the earth science classroom. In *Proceedings of the 1999 Annual Inter-national Conference of the Association for the Educatio of Teachers in Science*, (1232-1244). Greenville NC: Association for the Education of Teachers in Science.

Yost, N. (2001). Lights, Camera, Action: Videoconferencing in Kindergarten. In C. Crawford, et al. (Eds.), *Proceedings of Society for Information Technology and Teacher Education International Conference 2001*, (3173-3175). Chesapeake, VA: AACE.

In: Videoconferencing: Technology, Impact and Applications ISBN: 978-1-61668-285-9
Editor: Adam C. Rayler, pp. 111-122

Chapter 5

PROVIDING QUALITY END-TO-END VIDEOCONFERENCE SERVICES IN IP NETWORKS

Christos Bouras and Kostas Stamos*
Computer Engineering and Informatics Dept., University of Patras, Greece and Research Academic Computer Technology Institute, Greece

ABSTRACT

The development of quality and user satisfying videoconference applications at a low cost has been hindered in the past by, among other reasons, low quality connections and a difficulty for the average user to establish end-to-end connections without hassle. In this chapter, we describe some of the latest methods and developments that overcome these problems. In particular, we discuss ways to make establishment of universal videoconference connections easier by overcoming the problem of participants that reside behind NAT routers, by deploying new protocols at the application or even the network layer. An example of the former case is the development of establishment protocols specifically designed for this purpose, while an example for the latter case is the deployment of IPv6, which aims to make NAT deployments obsolete. Furthermore, we discuss efforts to provide targeted support for quality of experience in networks that support some kind of traffic classification through the usage of dynamic mechanisms and dynamic network re-configurations. Apart from describing the state of the art in the aforementioned areas, we provide detailed insight in specific research efforts that have taken place and assess their results in the overall user experience.

1. INTRODUCTION

Videoconferencing is the live exchange of audio and video information between people that find themselves in distant locations and are connected through a telecommunications channel. In order to place phone calls over the Internet, a number of open and closed VoIP

* Corresponding author: Email: {bouras, stamos}@cti.gr.

protocols have been developed. Videoconferencing offers cost and time advantages and through encryption can offer secure communication.

Traditionally, similar protocols have been used for the establishment of both VoIP and videoconference sessions, such as the H.323 and SIP (Session Initiation Protocol) standards. With the advent of cheap end user equipment and fast, reliable broadband connections, many other services such as instant messanging and VoIP clients have also added video to their conferencing capabilities (for example ICQ [67][68][66], Google Talk [68] and Skype [66]).

However VoIP and videoconferencing applications have not always been successful in offering a trouble-free experience to the user. Two of the main problems that users face have been the difficulty in setting up calls, especially when one or both of the communicating parties reside behind firewalls or NAT routers, and the difficulty to achieve satisfying quality under all circumstances, since Internet has been designed as a best-effort network and does not typically offer guaranteed service quality. The problem is that Internet traffic consists of flows generated by different applications, and all flows typically receive the same treatment from the network. However, real-time applications, such as VoIP and videoconferencing, are sensitive on parameters such as delay, packet loss or jitter. A worsening of these parameters results in measured inferior quality, as perceived by the users of a videoconference. This worsening can translate into interrupted audio or video, distorted speech and other artifacts that significantly reduce the user experience.

A lot of research and commercial effort has been spent in order to overcome the above problems and provide the opportunity for high quality end to end videoconferencing. Furthermore, the growth of the Internet has led the IETF to propose a new Internet Protocol, IPv6 [57], for a long-term solution to the address space shortage problem. Despite being standardized for over a decade, IPv6 has not yet enjoyed wide deployment. However, due to the oncoming IPv4 address space exhaustion [58], a transition to IPv6 seems inevitable in the near future. The abundance of IPv6 address space is also expected to obsolete NAT solutions that have hindered end-to-end videoconference deployment.

The rest of this chapter provides some background and describes some solutions that have been successfully deployed. In particular, the first part of the chapter deals with the establishment of end-to-end videoconferences and the solutions in order to traverse troublesome configurations such as NATs, while the second part of the chapter deals with the provisioning of quality service in a dynamic way, as videoconferencing participants enter and leave a call.

2. Establishing End-to-End Videoconferences

The scarcity of IPv4 addresses led to the development of the NAT (Network Address Translation) [56] technology for IP Masquerading, which is typically implemented in firewalls and routers in order to allow a single IP address to be used for multiple devices within a local network that wish to have network connectivity to the outside world. Each device within the local network is assigned a private IP address from a specified set of addresses that can not be globally routed (such as 192.168.x.x or 10.x.x.x), and the NAT router or firewall/router is responsible for translating the private IP addresses from the internal network to the routable public IP address. The translation is done by replacing the source address in all outgoing packets with the public IP address, while properly directing incoming packets (which all contain the same public IP address at the destination field) with the proper internal device. The NAT router keeps a translation table in order to identify which internal device has initiated which connection and properly forward incoming packets. The translation table keeps the basic data of each active connection, i.e. the destination address and the port number.

The problem with NAT is that the NAT router may not know where it should forward traffic originating outside the local network. Several rules and refinements have been proposed and implemented in various NAT devices that allow servers to be setup within the internal network, ranging from simple rules to forwards incoming connections to a specific machine, to port forwarding, where the connection port determines the receiving device.

However, NAT fundamentally alters the end-to-end concept upon which the Internet architecture is based, since the outside world thinks it is only communicating to the NAT router and not the actual internal machine. This breaks with assumptions made by protocols that expose IP information to upper layers, such as most standard video conference protocols. SIP and H.323 deal primarily with signalling and establishment of the connection, while audio and video traffic are transferred through a separate channel. Furthermore, these protocols utilize a number of ports that cannot be easily identified and dealt with by the NAT router/firewall. Therefore, the proper signalling sequence for establishing a call and then transferring audio and video data breaks because of the changes in the IP address contained in the exchanged packets.

The establishment of video conference calls between stations behind NAT configurations and the resulting user frustration and troubles has been a major hurdle in the widespread adoption of videoconferencing from casual users. Two main approaches can overcome this problem:

- Get rid of the need to use NAT
- Design protocols that can overcome hurdles related to NAT address translation.

For the first case, the most obvious solution is to migrate to the IPv6 protocol. IPv6 has been long touted as the inevitable replacement of IPv4, and because of its enormous increase in the address space, does away with the basic rationale for NAT usage. Lately it seems that widespread IPv6 adoption is coming closer. In this chapter, we detail the efforts to make a traditional videoconferencing protocol such as H.323 operate with IPv6 and therefore permit trouble-free end to end sessions [64].

The second approach has, for the time being, been more widely deployed in the form of new, often proprietary protocols such as Skype [66], which have become successful exactly because in large part they refrain from troubling the user with network issues and "just work". In a later section of this chapter we give a detailed description of Skype as a characteristic case. We also discuss NAT traversal using the STUN protocol [59], which in many cases allows SIP and H.323 connections to work behind NAT setups.

2.1. H.323 Voice and Video Network

2.1.1. H.323 calls

H.323 is an ITU recommendation, which defines a network architecture and the associated protocols necessary to voice and multi-media calls establishment. H.323 is defined for a packet-based network, and does not impose any network protocol, which can as well be IPv4 as IPv6 or IPX. H.323 architecture makes it possible to carry out direct calls between two multimedia phones connected on the Internet or a local area network. In this case, it is necessary to know the IP address of the called party. The main entities of an H.323 based video network are the following:

- End points: These are the H.323 clients, which are used by the end users. They can propose phone, video, fax, and application sharing functionalities.
- Gateways: Gateways can be used for the interconnection between different networks (for example an IP phone network and a traditional phone network).
- Gatekeepers: Gatekeeper makes it possible to be freed from the knowledge of called party IP address. It is then possible to call someone by his name. The gatekeeper is also able to manage the billing, and call filtering/authorization.
- Multipoint Control Units (MCUs): A Multipoint Control Unit (MCU) makes it possible to manage a conference of more than two end points. Each user connects to the MCU and then is able to discuss with all the other connected people.

Using a Gatekeeper or not, an H.323 phone call always follows the same logical order:

Optional registration to a gatekeeper (H.225)
Call setup, direct or thanks to gatekeeper (H.225)
Initial communication and capability exchange (H.245)
Establishment of audiovisual communication (H.245)
Call services (H.245, H.225)
Call termination (H.245, H.225)

The initial communication is done directly between the two phones. Those will thus have to use addresses understandable by both.

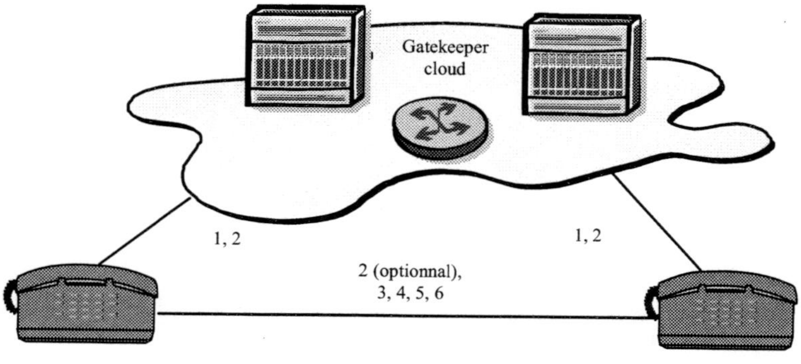

Figure 1. Call phases

	IPv4 server	IPv6 server
IPv4 client	Communicate using IPv4	Communicate using IPv4, server sees IPv4-mapped IPv6 address
IPv6 client	Can communicate using IPv4 if the IPv6 client uses an IPv4-mapped IPv6 address	Communicate using IPv6

Figure 2. Interoperability between IPv4 and IPv6 versions running on dual-stack hosts.

2.1.2. Impact of simultaneous IPv4 and IPv6 address on H.323

The current specification of the H.323 protocol makes it possible to manage calls independently of the network type. An H.323 transmission channel is defined by two communication terminations named endpoints: A source and a destination, which are defined by an address (IPv4, IPv6, IPX, etc.) and a Service Access Port (TCP/UDP Port).

In general, it is possible to achieve communication between IPv4 and IPv6 hosts, as shown on Figure 2.

It can however be a challenge to have an H.323 Call seamlessly using both IPv4 and IPv6 networks. When an IP phone software starts, it opens a communication port (Service Access Port) on which it could receive H.323 signaling. A dual-stack terminal can choose to open an IPv4 port, an IPv6 port or a dual protocol port. A dual port makes it able to receive both IPv4 and IPv6 call at the same time. On the other hand if it registers at a Gatekeeper, it has to make a choice and to specify an IPv4 or IPv6 address network.

As a gatekeeper maintains a list of the connected users (names or aliases), and corresponding endpoints (address and port), a dual terminal stack could be registered twice, once in IPv4 and another in IPv6.

It is possible to carry out an IPv4 and IPv6 mixed call between two dual stack terminals. Terminal A opens an IPv4 port and calls the terminal B, which opens an IPv6 port. The messages from A to B will be carried by the IPv6 network, and the messages from B to A will be in IPv4.

Problems arise if one wants to carry out a call between a pure IPv4 terminal and a pure IPv6 terminal. The use of mapped addresses is not possible, as H.225 messages at application layer clearly exchange IPv6 addresses.

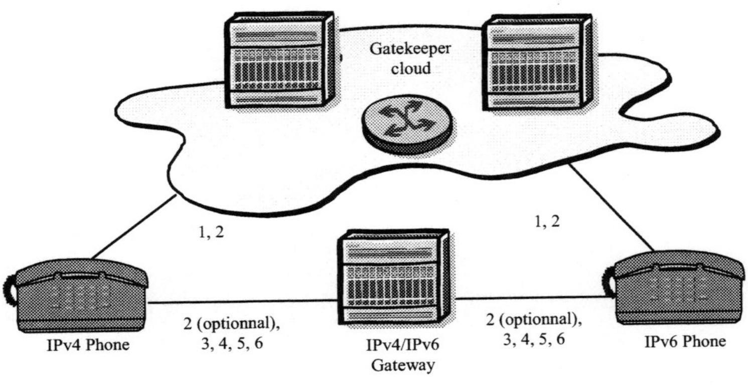

Figure 3. Call thanks to an IPv4/IPv6 Gateway.

A solution could be to use a gatekeeper connected to each network, able to detect this conflict and being used as Gateway for the transmission between the two networks.

Another solution could be to use at least one dual stack H323 client, able to detect an IPv4 only party, and sending the corresponding IPv4 address in H.225 messages instead of the mapped IPv6 address it is listening on.

Several efforts and experiments have been made with videoconferencing applications over IPv6 [64]. The problem of porting existing applications to IPv6 has been so far addressed by several researchers. A white paper by Microsoft [73] focuses on Windows applications, but at the same time offers some general guidelines that apply to any application for any operating system. In [74], the authors emphasize more on some general knowledge that a programmer must acquire before dealing with the problem of porting applications to IPv6, than on presenting step-by-step instructions. Furthermore, a number of research projects (6NET [75], Euro6IX [78], 6INIT [76], KAME [77]) have extensively investigated the migration effort and the benefits from IPv6, and have shared their experiences. The main objective has been to produce applications that can operate in both IPv4, IPv6 and dual-stack hosts. In general, these IPv6-enabled applications enable users to completely avoid the problematic NAT configurations and enjoy the benefits of global end-to-end connectivity. However, at the time being, despite very wide hardware and software support, at the network level IPv6 support is not universal among ISPs and therefore several intermediate solutions are required in order to achieve global IPv6 connectivity. Therefore, the user burden for proper NAT traversal configuration moves to the configuration of IPv6 transition mechanisms but is not eliminated. Nevertheless, most experts agree that IPv6 is the way of the future and as soon as it becomes the dominant Internet protocol, end-to-end communication is expected to become much easier.

2.2. NAT Traversal

Usually NAT configurations only permit TCP connections to be initiated from the hosts residing within the NAT network and not the outside Internet. Therefore, NAT traversal TCP connections can be quite problematic. Two commonly used options for overcoming this problem are relaying and connection reversal. The first method, relaying, is usually the only viable method if both communicating endpoints reside behind a NAT router. Relaying means that there needs to be a host with a global IP address acting as a relay between the two communicating endpoints. This relay host has TCP connections to both parties, and is then responsible for forwarding all traffic from one connection to the other. In most cases, the parties behind the NAT configurations need to have the relay associated with them before the TCP communication takes place. This means that in that way the address of the relay can be provided to each communicating party and not the actual address used by the communicating endpoint. TURN [63] is an example of a protocol that allocates and uses relay hosts. It is most useful for hosts that reside behind symmetric NAT routers or firewalls and intend to be on the receiving end of a connection to a single peer. TURN does not allow for users to run servers on well known ports if they are behind a NAT; it supports the connection of a user behind a NAT to only a single peer. In applications such as Skype, built on the peer-to-peer model, entities called supernodes act as relays and help the endpoints that reside behind NAT routers to communicate. In these cases, the allocation of relay hosts depends on the protocol used for the allocation of supernodes.

For UDP communications the type of the NAT configuration used is very important. A UDP packet sent through the NAT from the internal part of the network to the external Internet always creates a binding within the NAT between the internal and the external address and port pairs. Some NAT configurations forward packets sent from any address to the external address according to an active binding. However, some NATs only allow packets from the address where the original UDP packet that established the binding was sent to.

2.2.1. NAT traversal using stun

Stun is an Internet standards-track suite of methods, including a network protocol, used in NAT traversal for applications of real-time voice, video, messaging, and other interactive IP communications.

The Stun protocol allows applications operating through a network address translator (NAT) to discover the presence of a network address translator and to obtain the mapped (public) IP address (NAT address) and port number that the NAT has allocated for the application's User Datagram Protocol (UDP) connections to remote hosts. The protocol requires assistance from a 3rd-party network server (STUN server) located on the opposing (public) side of the NAT, usually the public Internet. The original version of the protocol also specified methods to ascertain the specific type of NAT, but those methods have been deprecated in the newer specification, because of the plethora of specific NAT implementation behavior in various networking equipment and the resulting intractability of the problem and the deficiencies of the method used.

Stun is not a self-contained NAT traversal solution applicable in all NAT deployment scenarios and does not work correctly with all of them. It is a tool among other methods, most notably Traversal Using Relay NAT (TURN) and Interactive Connectivity Establishment (ICE).

Stun does work with primarily three types: full cone NAT, restricted cone NAT, and port restricted cone NAT. In the cases of restricted cone or port restricted cone NATs, the client must send out a packet to the endpoint before the NAT will allow packets from the endpoint through to the client. STUN does not work with symmetric NAT (also known as bi-directional NAT) which is often found in the networks of large companies. Since the IP address of the STUN server is different than that of the endpoint, in the symmetric NAT case, the NAT mapping will be different for the STUN server than for an endpoint. TURN offers better results with symmetric NAT.

2.2.2. Proprietary protocols (Skype)

One of the most attractive features of Skype [66] and probably an important factor to its success, has been its ability to "just work", which means that users have no trouble making Skype calls even when they reside behind NAT routers and firewalls. Skype uses a proprietary, non-standardized and non-interoperable protocol that has not been made publicly available, however many important details about its operation are widely known. Specifically, it is based on a peer-to-peer model, and its architecture consists of two types of nodes, the supernodes and the ordinary nodes, as well as a central entity, the login server. A supernode is selected among ordinary Skype clients with the requirements of a public address and sufficicient CPU, memory and network bandwidth, in order to improve the overall availability of the system. Skype forms an overlay network: Supernodes maintain an overlay network among themselves, while ordinary nodes pick one or more supernodes to associate with. Supernodes are grouped into slots and slots are grouped into blocks. They relay communications to other clients behind a firewall. Ordinary nodes issue queries through the supernodes they are associated with. Each client builds and refreshes a list of reachable nodes known as the host cache. The host cache contains IP address and port numbers of supernodes. TCP is the protocol used for signaling.

It is believed [70] that Skype uses its own version of the STUN protocol in order to determine the type of NAT or firewall it is behind. Furthermore, it seems that there is no server containing this information, because experiments have not found this type of network exchange of kind of information. The information about the type of NAT or firewall is dynamically determined and then locally stored and periodically refreshed in the Windows registry. A Skype client can traverse NAT routers and firewalls and needs no explicit NAT or firewall traversal server. The main characteristics that enable this are:

- The random selection of sender and listener ports by Skype: The client randomly chooses a port number upon installation, which can be configured in its connection dialog box.
- The usage of TCP as the voice streaming protocol when UDP can not be used: Specifically, Skype is based on the detection of the type of NAT or firewall. If both communicating endpoints use global IP address, then media traffic flows directly between them using UDP. If either one or both the endpoints are behind port-restricted NAT, they sent voice traffic to a node that acts as media proxy using UDP. If both endpoints reside behind port-restricted NAT and UDP-restricted firewalls, then the TCP protocol is used for transferring voice traffic over from another online Skype node.
- The peer-to-peer nature of the Skype network.
- The usage of the UDP hole punching technique: After an initial contact with the intermediate host in the public address space, the endpoints switch to direct communication hoping that the NAT devices will keep the states despite the packets coming from a different host.

Although Skype has achieved admirable NAT traversal in most cases and worldwide adoption, its lack of interoperability and lack of openness about the details of its architecture can be a limiting factor especially when interoperability with standard protocols and solution from different vendors or legacy equipment is a requirement.

3. Providing Targeted QoS

Setting up the videoconference successfully only solves half the problem, as many users find the experience to be underwhelming in terms of perceived quality. Videoconferencing presents a difficult challenge for best effort networks, because it requires a large amount of bandwidth and its real-time nature imposes strict requirements in terms of network parameters such as delay and jitter. Many networks operate with a constant heavy load in order to maximize resource utilization, which increases buffer queues in the network's routing devices and degrades the user experience because of lost or delayed packets.

Therefore, this section deals with the issue of providing strict guarantees and ensuring high quality videoconferences in networks that implement the logic suitable to enforce these guarantees.

3.1. Dynamic Qos Provisioning on the Fly

This section describes a solution for dynamic QoS provisioning that was experimentally implemented at the greek research network GRNET [65]. GRNET provides network services for research and academic organizations (institutes and universities) in Greece, which are its clients.

Figure 4 shows a typical setup for a videoconference between three users. GRNET uses a central MCU in order to facilitate videoconferences with multiple participants. All users have to connect to this MCU, with which they exchange traffic. The MCU is responsible for receiving audio and video streams from the participants and then multiplexing the streams and transmitting it back to all the participants. This means that whenever the network experiences congestion, the quality of the whole videoconference can be affected. This applies even in the case where a single network link is congested. If the packets originating from a single participant traverse the congested link, all the participating users will experience degraded quality, because the audio and video stream will be multiplexed by the MCU and sent to everybody.

Grnet does implement Quality of Service (QoS), which guarantees traffic parameters for high quality connections. The architecture model is based on DiffServ [60] which is based on the concept of traffic classes. Each traffic class is mapped to a Per-Hop Behavior (PHB), and PHBs are implemented at routers by means of queuing and scheduling at congestion points, where queues are formed. By mapping different traffic types into different PHBs, routers are able to ensure service guarantees. The provisioning model at GRNET requires only policing at the network perimeter, while core routers implement priority queuing mechanism, and admission control based solely on the availability of IP premium bandwidth at the access links. User requests for QoS reservations are made through a management tool developed by GRNET, called ANStool [62].

The ANStool therefore allows the creation of permanent (or long-term) QoS configurations that would treat traffic from the MCU to every destination as priority traffic and treat traffic from every source to the MCU as priority traffic. This approach is however not attractive in the current setting of dynamic and short-term videoconferences. First of all, such a static configuration should be applied at the network perimeter to make sure traffic is marked as priority throughout its whole journey though the network. This means that there is no way to know whether incoming traffic corresponds to one or more users from the same source (client institute to GRNET), and therefore no way to calculate how much traffic should be prioritized. Allowing the maximum possible traffic would mean that the admission control architecture of the QoS provisioning service would be completely bypassed.

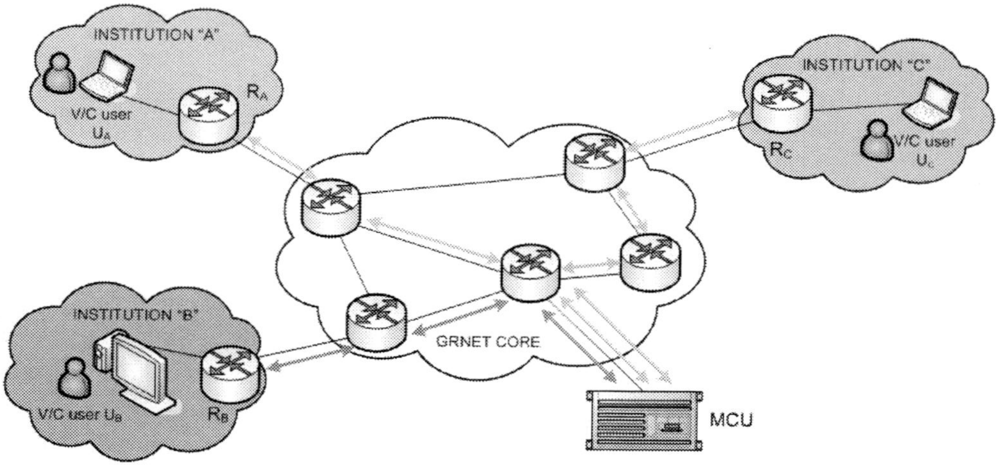

Figure 4. Typical videoconference setup [65]

A related problem is that the bypassing of admission control leads to the possibility of DOS (denial of service) attacks, by injecting malevolent traffic towards the MCU (illegitimate RTP packets cannot easily told apart and filtered out by ordinary router mechanisms such as access-lists). This traffic would be treated with priority throughout the network, with the danger of summing up to high bandwidths and disrupting the normal operation of the network, the MCU or both.

Therefore, GRNET decided to implement a mechanism that creates, propagates, and installs on-the-fly rules for admission control and marking of videoconference traffic at the network perimeter (where traffic is marked and policed for conformance to contracted-to traffic profiles). In this manner, only traffic from legitimate users is prioritized. No other users may inject priority traffic in the network. Moreover, this method performs admission control, so that if multiple users from a given institution participate in a videoconference, it is possible that only some of them will be allowed to inject priority traffic, while others will not.

The mechanism operates by taking into account a maximum amount of bandwidth that each client institute has allocated for videoconferencing. Once every minute, a script queries the MCU and retrieves the IP addresses of all users participating at selected videoconferencing sessions. The script then checks which of these participants are authorized for prioritized traffic and queries the network routers for their FIB (Forwarding Information Base) paths towards the QoS-authorized participants. The path information is stored in a database, where it is retrieved by a dedicated router daemon, that then advertises it to all routers as /32 addresses over a specially-crafted BGP (Border Gateway Protocol) [61] session. Therefore, the database serves as the intermediate between the script and the routing daemon, and contains an up-to-date view of the participants that require preferential (QoS) treatment in a dynamic way. The /32 routes advertised to the routers take precedence over the normal BGP route and the routers are configured to insert these routes in their FIB with an Expedited Forwarding tag, for QoS treatment. This configuration at GRNET is currently Cisco-specific, called by Cisco QoS Propagated Policy via BGP (QPPB). This tag signals that matching packets should be classified as priority traffic and sent to a low latency queue.

More technical and implementation details for this service can be found at [65].

4. CONCLUSION

We have seen in this chapter how the emergence of quality and user satisfying videoconference applications can be achieved by advanced techniques for overcoming past problems of NAT penetration and network quality guarantees. We have described recent methods for the establishment of universal quality videoconference connections and we have discussed how this application field will be affected by the upcoming migration of the Internet to IPv6.

Videoconference applications seem to have overcome many of the obstacles that hindered their widespread adoption in the past. Future problems might shift from technical challenges to challenges related to the end user perception of the services, for example the preference of audio communication over video and audio communication, usability features as well as the level of compatibility and interoperability of future videoconferencing solutions.

5. REFERENCES

[56] RFC, (20001). 3022, "*Traditional IP Network Address Translator* (Traditional NAT)", P. Srisuresh, & K. Egevang, (January).

[57] RFC, (1998). 2460, "*Internet Protocol*, Version 6 (IPv6) Specification", S. Deering, & R. Hinden (December).

[58] American Registry for Internet Numbers, "*ARIN Board Advises Internet Community on Migration to IPv6*", May 2007, http://lists.arin.net/pipermail/info/2007-May/000111.html

[59] RFC, (2003). 3489, "*STUN - Simple Traversal of User Datagram Protocol (UDP) Through Network Address Translators (NATs)*", J., Rosenberg, J., Weinberger, C. Huitema, & R. Mahy, The Internet Society (March)

[60] RFC, (1998). 2475, "*An Architecture for Differentiated Services*", S., Blake, D., Black, M., Carlson, E., Davies, Z. & Wang, W. Weiss, (December).

[61] RFC, (2006). 4271, "*A Border Gateway Protocol 4 (BGP-4)*", Y., Rekhter, T. Li, & S. Hares, (January)

[62] Varvitsiotis, A., Siris, V., Primpas, D., Fotiadis, G., Liakopoulos, A. & Bouras, C. (2005). "Techniques for DiffServ - based QoS in Hierarchically Federated MAN Networks - the GRNET Case", *The 14th IEEE Workshop on Local and Metropolitan Area Networks* (LANMAN 2005), Chania. Island of Crete, Greece, 18-21 September.

[63] Internet-Draft, (2005). "*Traversal Using Relay NAT (TURN)*", Rosenberg, J., Mahy, R. & Huitema, C. September, (work in progress)

[64] Bouras, C., Josset, S., Gkamas, A. & Stamos, K. (2003). "Adding IPv6 support to H323 Gnomemeeting/openH323 port" *11th International Conference on Software*, Telecommunications and Computer Networks (SoftCOM 2003), Croatia, Italy, October 7 - 10 2003, 458-462.

[65] http://rts.grnet.gr/vc-qos.php

[66] Skype software, www.skype.com.

[67] ICQ, http://www.icq.com/

[68] Google Talk, http://www.google.com/talk/

[69] Desclaux, F. & Kortchinsky, K. (2006). "*Vanilla Skype*" (parts 1 and 2), June.

[70] Baset, S. A. & Schulzrinne, H. G. (2006). "*An analysis of the skype peer-to-peer internet telephony protocol*," in INFOCOM 2006. 25th IEEE International Conference on Computer Communications. Proceedings, 1-11.

[71] Sven Ehlert and Sandrine Petgang, "*Analysis and signature of skype voip session traffic*", Fraunhofer FOKUS Technical report NGNI-SKYPE-06b, 2006.

[72] Markus Isomäki,. "*Peer-to-Peer Communication Services in the Internet*", Nokia Research Center, 2003.

[73] "*Adding IPv6 capability to Windows Socket Applications*", Microsoft Corporation.

[74] "*Porting Networking Applications to the IPv6 APIs*", Sun Microsystems.

[75] 6NET project, http://www.sixnet.org.

[76] 6INIT project, http://www.6init.org/presentations.html.

[77] KAME project, http://www.kame.net/

[78] Euro6IX project, http://www.euro6ix.net

In: Videoconferencing: Technology, Impact and Applications ISBN: 978-1-61668-285-9
Editor: Adam C. Rayler, pp. 123-131

Chapter 6

Utilizing Web-Based Videoconferencing to Improve the Continuum of Care in Long-Term Outpatient Substance Abuse Treatment

Van L. King, Robert K. Brooner and Michael S. Kidorf
The Johns Hopkins School of Medicine and Addiction Treatment Services at Hopkins Bayview, Baltimore, Maryland, USA

Abstract

Web-based videoconferencing technologies hold far-reaching potential to increase and improve delivery of evidence-based substance abuse treatment interventions. A review of the literature reveals very few reports on the efficacy of web-based videoconferencing intervention strategies for treating substance use disorder, and many of those have been descriptive reports that do not meet the requirements of establishing effective and replicable evidence-based care. Comparatively more studies have reported on the use of computer-administered pre-programmed treatment and education modules. These approaches can be very helpful, but do not expand the availability of flexible and professional counseling services for patients in long-term therapy relationships that can be realized with web-based videoconferencing. One of the next important advances in the substance abuse treatment continuum is to evaluate how web-based videoconferencing can help engage patients with serious and chronic problems in the long-term outpatient treatment process. Patients requiring long-term treatment often require responsive and flexible levels of counseling to manage symptoms that fluctuate greatly in scope and intensity over time. Unfortunately, the overall effectiveness of long-term therapy is often seriously hindered by treatment demands to on-site scheduled counseling and psychosocial services that often continue for many years. These demands reduce utilization of therapy services with resulting negative impacts on treatment outcome. Web-based videoconferencing service delivery provides a remarkable opportunity to reduce some of the inconvenience of counseling and so facilitate the long-term treatment process while at the same time allowing pursuit of other responsibilities (e.g., child care) or treatment goals (e.g., employment) essential to rehabilitation. A recent study by our research group demonstrated both increased satisfaction and improved attendance in patients receiving long-term substance abuse treatment who were assigned to web-based counseling versus on-site counseling, with no differences in treatment outcomes. However, much more work is needed to establish how best to use this technology since the great majority of published studies have focused on individuals with mild to moderate problems and/ or on brief interventions. Using web-based videoconferencing to expand the continuum of care holds great promise to improve treatment satisfaction, adherence and retention for the many patients with chronic substance use problems requiring long-term outpatient substance abuse treatment.

INTRODUCTION

Web-based videoconferencing technologies hold far-reaching potential to increase and improve the delivery of long-term evidence-based substance abuse interventions because they permit substance users to access "real time" professional therapy from the convenience of their homes. However, though many web sites offer substance abuse treatment to persons with some types of substance use disorders (such as nicotine and alcohol), there is clearly a major gap in the literature examining the potential benefits of Internet-based therapy for illicit drugs of abuse that often benefit from more intensive, long-term treatment (Griffiths, 2005). A review of the literature shows that most of the e-therapy and telemedicine studies published to date evaluate psychiatric and substance abuse treatments with minimal therapist contact (Carlbring, et al., 2007) or brief, self-help interventions (Koski-Jannes et al., 2007; Ruggiero et al., 2006; Hester et al., 2005; Walters et al., 2007; Pull, 2006; Hailey et al., 2008). These interventions were conceived as a way to more efficiently reach larger numbers of people who might not otherwise receive psychiatric or substance abuse education or treatment, or as a type of "pre-therapy" to prepare and motivate concerned individuals to access more intensive treatment after the intervention (Griffiths, 2005). These approaches can be very helpful, but do not expand the availability of flexible and professional counseling services for patients already in treatment that can be facilitated by the use of web-based videoconferencing service delivery platforms.

Very few reports evaluate more intensive, professionalized interventions such as web-based individual or group therapy (Frueh et al., 2007; Marziali et al., 2005), and even fewer have specifically evaluated these interventions for substance abuse treatment (Frueh et al., 2005; King et al., 2009). Indeed, a recent meta-analysis showed that short-term web-based interventions were better attended and more effective when accompanied by direct therapist involvement (Spek et al., 2007). This finding has profound implications for the effective development and employment of these technologies in populations with more complex and serious disorders that will benefit from more direct professional involvement because web-based videoconferencing is a very effective means to increase direct therapist contact. Researchers are only beginning to evaluate web-based videoconferencing for these purposes.

Fixed site, point-to-point videoconferencing communication has been demonstrated to be as effective as on-site services for psychiatric evaluations and follow-ups in multiple studies (Hyler et al., 2005; Myers et al., 2007; Lexcen et al., 2006; O'Reilly et al., 2007). It has also been shown to work very effectively to administer psychotherapy to incarcerated patients with some significant logistic advantages for treatment providers, considering the difficulty with accessing patients in the correctional setting (Morgan et al., 2008; Antonacci et al., 2008). There is even some indication that patients can be more revealing earlier in the psychotherapeutic process using videoconferencing compared to in-person interaction (Day & Schneider, 2002). Similarly, a small but growing number of studies have shown good results using web-based videoconferencing as a platform for delivering circumscribed psychiatric treatment (Khasanshina et al., 2006) and psychotherapy (Frueh et al., 2005; Frueh et al., 2007; King et al., 2009). King et al. (2009) evaluated a web-based videoconferencing platform (eGetgoing, CRC Health Group, Inc.) to deliver intensified group substance abuse counseling to long-term outpatients responding poorly to less intensive services. Outcomes in this randomized and controlled trial were both good and comparable across the control and web-based conditions, but patients uniformly preferred the videoconferencing condition. However, no published studies to date examine web-based videoconferencing to administer long-term routine or intensified psychotherapies for substance abuse or any other psychiatric problems. One of the next important advances in the substance abuse treatment continuum is to evaluate how web-based videoconferencing can help improve engagement and delivery of care to patients with serious and chronic problems in the long-term outpatient treatment setting.

The Burdens of Long-Term Treatments

Patients with serious and chronic substance use disorders often benefit from long-term treatment. These treatments require responsive and flexible levels of counseling to manage symptoms that fluctuate greatly in scope and intensity over time. Unfortunately, the overall effectiveness of therapy is often seriously hindered by treatment demands to on-site scheduled services that may continue for many years. These demands reduce utilization of therapy services with resulting negative impact on treatment outcome (Crits-Christoph and Siqueland, 1996; Kidorf et al., 2006). Unfortunately, problems with poor adherence to treatments for chronic medical conditions are common. The World Health Organization has identified adherence as one of the most important problems affecting treatment effectiveness, and improving treatment adherence is perhaps the best way to improve most treatment outcomes (Sabate, 2003).

Opioid Treatment Programs (OTPs) are the prototypical long-term outpatient substance abuse treatment. OTPs are widely regarded as effective settings for the care of people with chronic opioid dependence, as measured by reductions in opioid use, infectious disease risk behaviors, the spread of HIV, and criminal behavior (McLellan et al., 2000). Patients receive daily methadone to reduce and eliminate illicit opioid use and concomitant counseling therapy to help the psychosocial rehabilitation process. There are a large number of OTPs in the US (over 1,200) treating over 250,000 patients. The benefits of methadone and other opioid agonists can be significantly improved by routine counseling and other enhanced psychotherapies (Moos, 2007). Unfortunately, poor attendance by patients to even routine low intensity counseling schedules limits the benefits of this service. For example, in his review of the literature, Kidorf et al. (2006) found that less than 50% of scheduled counseling sessions are delivered to patients in many substance abuse treatment settings. Because of the long-term need for counseling in OTPs, they are exceptionally well-suited sites for developing web-based videoconferencing technology to support these treatment needs.

VIDEOCONFERENCING CAN REDUCE THE BURDENS OF TREATMENT

OTPs may be excellent settings to test the impact of web-based videoconferencing platforms, but this population and setting has been largely neglected in studies utilizing web-based treatment approaches. Poor attendance is common in these programs, so web-based delivery of routine therapy might be one way to improve participation. For example, web-based delivery of counseling services could expand the limited range of hours programs now use to provide counseling. These services could occur outside of the regularly scheduled times for on-site appointments, and counselors could then focus their attention on more unstable patients during the times the clinic was open for on-site appointments. In addition, the delivery of group counseling can be more convenient because it is orchestrated from the counselor's office and extends to patients participating in the comfort and privacy of their own homes without the necessity of on-site attendance in a large office space.

Employment, child-care, medical, and other time consuming appointments interfere with attendance to scheduled counseling services and frequently reduce service utilization, which is often associated with poor outcomes in multiple areas of functioning. Web-based videoconferencing platforms provide an exciting opportunity to reduce some of the inconvenience of treatment and allow patients more time to pursue the other responsibilities and treatment goals essential to a more complete rehabilitation. For example, working patients could receive counseling services from home following a day of work, or could utilize expanded late or even weekend hours pre-arranged by the clinical staff. Mothers with small children could engage in counseling at home. A web-based service delivery model could also accommodate patients from a wider geographic area. Cost factors also appear to be associated with poor adherence. Community-based treatment programs often treat individuals with limited financial resources, many of whom are supported by public assistance programs. Internet web-based delivery systems have strong potential to reduce the number of weekly (and even daily) trips patients are required to make to the treatment program, thereby reducing overall transportation costs while concurrently improving rates of adherence. It should be noted here that despite the limited financial resources of this treatment population, many of them have good private access to computers and the Internet. A recent survey conducted at our community-based program (n = 160) found that 50% had a home computer with Internet access.

In addition to offering direct and responsive professional contact, videoconferencing is very flexible and could be used to deliver a variety of evidenced-based treatment approaches. This makes videoconferencing particularly attractive for use with chronic substance dependent individuals that often have a number of comorbid psychiatric and psychosocial problems. At least one study has already shown that psychiatric consultation over the web can be quite effective (Khasanshina et al., 2006). And finally, web-based videoconferencing can be used to offer professional care over long periods of time, which is ideal for individuals with chronic and relapsing disorders like substance dependence whose symptoms fluctuate in range and intensity over the course of the disorder.

A final potential benefactor of this new technology is the community in which the OTP or other long-term treatment program is located. It's well known that some of the most vigorous opposition to the establishment and operation of community-based treatment programs come from community members concerned about the volume of highly symptomatic patients visiting the clinic daily with attendant disruption of the day-to-day activities of the community. Web-based counseling approaches hold considerable potential to lessen many of these problems.

Intensified Group Counseling Utilizing Videoconferencing

A recent study demonstrated both increased satisfaction and improved attendance to an intensified substance abuse group therapy treatment. This report is the only published study using this technological approach to expand the continuum of care in an OTP (King et al., 2009). After informed written consent was obtained, subjects responding poorly to standard (lower intensity) counseling schedules (n = 37) were randomly assigned to a more intensive schedule of group counseling utilizing either: 1) a standard on-site delivery condition, or 2) a web-based videoconferencing service delivery condition. Participants assigned to the standard care condition attended the clinic to receive their on-site intensified counseling sessions, while those in the videoconferencing service condition attended the same set of intensified sessions via the Internet. The only difference between the study conditions was the treatment delivery platform.

While participants in both conditions had a good, and often comparable, response to the intensified counseling schedule, participants assigned to the videoconferencing service delivery condition had a substantially higher rate of adherence to the group counseling sessions (M = 92% vs. 76%, p = .07), although the difference was not statistically significant due to the small sample size. Adherence rates of over 90% are rarely observed in the substance abuse treatment literature. Another important finding was that no differences were observed in urine tests positive for drug use (M = 37% videoconferencing vs. 42% routine, p = ns), indicating that evidenced-based counseling sessions were comparably effective whether delivered on-site or via the Internet. Participants receiving web-based therapy consistently rated their satisfaction with videoconferencing service delivery higher than that reported by participants in the on-site condition. In addition, all of the participants exposed to the web-based therapy completed a separate questionnaire about their preference for videoconferencing service delivery versus on-site care should they require another episode of intensified care in the future, and 100% reported a strong preference for the Internet delivery condition. Taken together, these results provide good evidence of the feasibility of incorporating web-based technology within OTPs and other community-based treatment clinics and highlights some of the benefits that this technology can bring to efforts to expand the access and scope of substance abuse treatment services.

Future Research

Web-based videoconferencing service delivery provides a remarkable opportunity to reduce some of the burdens of treatment and so facilitate the long-term treatment process. However, much more work is needed to establish how best to use this technology. Since few studies have been conducted with chronic substance users, the existing literature does not generalize easily to this critical patient population. In addition, most reports have evaluated these technologies as stand-alone interventions rather than as part of a continuum of services offered at fixed-site treatment programs. One way of expanding services in these fixed-site settings is to use web-based technologies as a way to intensify treatment to unstable patients (King et al., 2009; Carroll et al., 2008). However, web-based services may also be conceptualized as a less restrictive component of care within a broader adaptive continuum of service delivery, reserved for patients that have attained clinical stability in regard to drug use and other program objectives. Web-based videoconferencing could not only be used to deliver long-term treatment to OTP patients, but could also be used to treat patients with chronic stimulant, alcohol or marijuana dependence who are finished with the acute phase of treatment yet require significant periods of counseling and monitoring to establish long-term drug abstinence (McKay, 2005). This area of research is considerably underdeveloped, in some measure due to the difficulty in keeping patients in treatment over extended periods of time. Videoconferencing technology offers flexibility and convenience that could be the critical components keeping a patient connected to and participating in ongoing treatment services.

The positive features of web-based videoconferencing platforms are also highly relevant to the growing number of medical practices offering long-term opioid agonist treatment with buprenorphine (Ikelheimer, 2008). Physicians that offer office-based buprenorphine treatment typically rely on low intensity counseling schedules, so web-based delivery of counseling services might be acceptable to patients and providers as a way to intensify treatment as necessary. A web-based program might attract patients who would otherwise not attend additional counseling, and complementary treatment between the physician and the web-based therapist would likely improve treatment response. Future studies focusing on buprenorphine (or methadone) office-based treatment could evaluate how videoconferencing technology could enhance service delivery by providing more convenient access to treatment staff overseeing both the medication and psychosocial components of the interventions (King et al., 2006; Stoller et al., 2006).

Finally, available studies have not adequately explored some unintended but possible effects of relying on web-based delivery of routine care services that could affect acceptance and transportability to community-based programs. For example, variables such as the patient-therapist relationship might be adversely affected by over-reliance on computer or web-based standards of care. Though existing studies have not demonstrated consistent and significant problems with the therapeutic relationship using videoconferencing technologies, there are still significant questions about the stability and characteristics of therapeutic relationships over longer periods of time that may not be apparent in shorter time frames. Only more extended evaluations can determine how well web-based videoconferencing can act as a service delivery platform for long-term counseling relationships.

CONCLUSION

In summary, much of the small amount of research available on the use of computer-supported technologies to provide greater and more convenient access to substance abuse services is descriptive in nature, and most of it is based on pre-programmed computer learning modules versus real-time web-based interaction between patients and therapy staff.

One of the few exceptions to this is the study by King et al. (2009) that clearly supports both the feasibility and effectiveness of web-based counseling services for an OTP population. The inherent flexibility of web-based videoconferencing approaches makes them ideally suited for increasing and improving the delivery of counseling service to patients in long-term treatment settings. That study also shows that even low-income, relatively unstable patients, had access to the Internet. Web-based delivery systems have good potential to improve utilization of and adherence to routine and intensified counseling schedules by reducing inconvenience and treatment-related time and travel costs with fewer clinic space demands and shorter hours of on-site service delivery. Additional research employing randomized and controlled evaluations of the benefits and potential limitations of these technologies, and potential solutions to the challenges of routine use of these technologies, would advance the field.

REFERENCES

Antonacci, D. J., Bloch, R. M., Saeed, S. A., Yildirim, Y. & Talley, J. (2008). Empirical evidence on the use and effectiveness of telepsychiatry via videoconferencing: Implications for forensic and correctional psychiatry. *Behavioral Sciences and the Law, 26*, 253-269.

Carlbring, P., Gunnarsdottir, M., Hedensjo, L., Andersson, G., Ekselius, L. & Furmark, T. (2007). Treatment of social phobia: randomized trial of internet-delivered cognitive-behavioural therapy with telephone support. *British Journal of Psychiatry, 190*, 123-128.

Carroll, K. M., Ball, S. A., Martino, S., Nich, C., Babuscio, T. A., Nuro, K. F., Gordon, M. A., Portnoy, G. A. & Rounsaville, B. J. (2008). Computer-assisted delivery of cognitive-behavioral therapy for addiction: a randomized trial of CBT4CBT. *American Journal of Psychiatry, 165*, 881-888.

Crits-Christoph, P., Siqueland, L. & Blaine, J., et al. (1999). Psychosocial treatments for cocaine dependence. *Archives of General Psychiatry, 56*, 493-502.

Day, S. X. & Schneider, P. L. (2002). Psychotherapy using distance technology: a comparison of face-to-face, video, and audio treatment. *Journal of Counseling Psychology, 49*, 499-503.

Frueh, B. C., Henderson, S. & Myrick, H. (2005). Telehealth service delivery for persons with alcoholism. *Journal of Telemedicine and Telecare, 11*, 372-375.

Frueh, B. C., Monnier, J., Grubaugh, A. L., Elhai, J. D., Yim, E. & Knapp, R. (2007). Therapist adherence and competence with manualized cognitive-behavioral therapy for PTSD delivered via videoconferencing technology. *Behavior Modification, 31*, 856-866.

Griffiths, M. (2005). Online therapy for addictive behaviors. *Cyberpsychology & Behavior, 8*, 555-561.

Hailey, D., Roine, R. & Ohinmaa, A. (2008). The effectiveness of telemental health applications: a review. *Canadian Journal of Psychiatry, 53*, 769-778.

Hester, R. K., Squires, D. D. & Delaney, H. D. (2005). The Drinker's Check-up: 12-month outcomes of a controlled trial of a stand-alone software program for problem drinkers. *Journal of Substance Abuse Treatment, 28*, 159-169.

Hyler, S. E., Gangure, D. P. & Batchelder, S. T. (2005). Can telepsychiatry replace in-person psychiatric assessments? A review and meta-analysis of comparison studies. *CNS Spectrums*, *10*, 403-413.

Ikelheimer, D. M. (2008). Treatment of opioid dependence via home-based telepsychiatry. *Psychiatric Services*, *59*, 1218-1219.

Khasanshina, E. V., Wolfe, W. L., Emerson, E. N. & Stachura, M. E. (2006). Counseling center-based tele-mental health for students at a rural university. *Telemedicine and e-Health*, *14*, 35-41.

Kidorf, M., King, V. L. & Brooner, R. K. (2006). Counseling and psychosocial services, in Strain, EC, and Stitzer, ML (Eds.), *The Treatment of Opioid Dependence*, Baltimore, MD: Johns Hopkins University Press.

King, V. L., Kidorf, M. S., Stoller, K. B., Schwartz, R., Kolodner, K. & Brooner, R. K. (2006). A 12-month controlled trial of methadone medical maintenance integrated into an adaptive treatment model. *Journal of Substance Abuse Treatment*, *31*, 385-393.

King. V. L., Stoller, K. B., Kidorf, M. S., Kindbom, K., Hursh, S., Brady, T. & Brooner, R. K. (2009). Assessing the effectiveness of an Internet-based videoconferencing platform for delivering intensified substance abuse counseling. *Journal of Substance Abuse Treatment*, *36*, 331-338.

Koski-Jannes, A., Cunningham, J. A., Tolonen, K. & Bothas, H. (2007). Internet-based self-assessment of drinking – 3-month follow-up data. *Addictive Behaviors*, *32*, 533-542.

Lexcen, F. J., Hawk, G. L., Herrick, S. & Blank, M. B. (2006). Use of video conferencing for psychiatric and forensic evaluations. *Psychiatric Services*, *57*, 713-715.

Marziali, E., Donahue, P. & Crossin, G. (2005). Caring for others: Internet health care support intervention for family caregivers of persons with Alzheimer's, stroke, or Parkinson's Disease. *Families in Society*, *86*, 375-383.

McKay, J. R. (2005). Is there a case for extended interventions for alcohol and drug use disorders? *Addiction*, *100*, 1594-1610.

McLellan, A. T., Lewis, D. C., O'Brien, C. P. & Kleber, H. D. (2000). Drug dependence, a chronic medical illness. *JAMA*, *284*, 1689-1695.

Moos, R. H. (2007). Theory-based active ingredients of effective treatments for substance use disorders. *Drug and Alcohol Dependence*, *88*, 109-121.

Morgan, R. D., Patrick, A. R. & Magaletta, P. R. (2008). Does the use of telemental health alter the treatment experience? Inmate's perceptions of telemental health versus face-to-face treatment modalities. *Journal of Consulting and Clinical Psychology*, *76*, 158-162.

Myers, K. M., Valentine, J. M. & Melzer, S. M. (2007). Feasibility, acceptability, and sustainability of telepsychiatry for children and adolescents. *Psychiatric Services*, *58*, 1493-1496.

O'Reilly, R., Bishop, J., Maddox, K., Hutchinson, L., Fisman, M. & Takhar, J. (2007). Is telepsychiatry equivalent to face-to-face psychiatry? Results from a randomized controlled equivalence trial. *Psychiatric Services*, *58*, 836-843.

Petry, N. M. (2000). A comprehensive guide to the application of contingency management procedures in clinical settings. *Drug and Alcohol Dependence*, *58*, 9-25.

Pull, C. (2006). Self-help internet interventions for mental disorders. *Current Opinions in Psychiatry*, *19*, 50-53.

Ruggiero, K., Resnick, H., Acierno, R., Carpenter, M., Kilpatrick, D., Coffey, S., Ruscio, A., Stephens, R., Stasiewicz, P., Roffman, R., Bucuvalas, M. & Galea, S. (2006). Internet-based intervention for mental health and substance use problems in disaster-affected populations: a pilot feasibility study. *Behavior Therapy*, *37*, 190-205.

Sabate, Eduardo. (2003). Adherence to long-term therapies: evidence for action. World Health Organization, 196pgs. ISBN # 92241545992.

Spek, V., Cuijpers, P., Nyklicek, I., Riper, H., Keyzer, J. & Pop, V. (2007). Internet-based cognitive behaviour therapy for symptoms of depression and anxiety: a meta-analysis. *Psychological Medicine*, *37*, 319-328.

Stitzer, M. L., Petry, N. & Silverman, K. (2006). Contingency management therapies. In E. C. Strain, & M. L. Stitzer (Eds.), *The Treatment of Opioid Dependence.* (151-177). Baltimore: Johns Hopkins University Press.

Stoller, K. B., King, V. L., Clark, M. R. & Brooner, R. K. (2006). Office-Based Buprenorphine Maintenance: setting the stage for success. *Johns Hopkins Advanced Studies in Medicine*, *6*, 124-134.

Walters, S. T., Vader, A. M. & Harris, T. R. (2007). A controlled trial of web-based feedback for heavy drinking college students. *Prevention Science*, *8*, 83-88.

In: Videoconferencing: Technology, Impact and Applications ISBN: 978-1-61668-285-9
Editor: Adam C. Rayler, pp. 133-177

Chapter 7

INTERACTIVE VIDEOCONFERENCING (IVC) AS A CRUCIAL FACTOR IN K-12 EDUCATION: TOWARDS A CONSTRUCTIVISM IVC PEDAGOGY MODEL UNDER A CROSS-CURRICULAR THEMATIC APPROACH

Panagiotes S. Anastasiades*
University of Crete, Rethymnon, Greece

ABSTRACT

The advanced learning internet technologies indicated a need for methodologies, tools and pedagogies that support the introduction of synchronous e- learning applications and especially Interactive Videoconferencing (IVC) into the whole learning environment. The objective of this chapter is to present the theoretical background and the methodology that have been developed, to implement a pedagogical model for IVC instruction. The proposed pedagogical model constitutes the results experienced from an IVC project in the University of Crete during the last seven years, under the title `ODUSSEAS', which took place between eight hundred students and thirty teachers in eighteen elementary schools in Greece and Cyprus.

Keywords: Interactive Videoconference, Videoconferencing pedagogy, Virtual Classroom, Collaborative activities by distance

* Associate Professor, University of Crete, Department of Education, Greece. Email: panas@edc.uoc.gr
Home Page: http://www.edc.uoc.gr/~panas/

1. THE NEW LEARNING ENVIRONMENT - INTRODUCTION

A new digital continent is thus born in which concepts like borders, geographical distances, movement or means of transportation, no longer matter. The education sector is the Knowledge Society's main priority, one of the most important portals of the new digital age (Berge & Collins, 1995; Anastasiades, 2002a).

The ICTs' under specific conditions, are expected to provoke numerous, important changes in the contemporary educational environment, as it has already occurred in the wider productive and socio-cultural environment of our times. The communication between students and learners is not dependent on geographical distance but rather on their ability to access the network (Anastasiades, 2002b; Burge, 2001; Fitzgerald, 2000). The diverse and powerful ICT are not only called out to replace the traditional teaching methodologies, but also to enrich and expand them, as well as offer new possibilities of upgrading the traditional educational environment through an innovative, transformative pedagogical approach, which will aim at central changes in all the parameters of the educational procedure (Raptis&Rapti, 2004).

According to Anastasiades (2003a) the new nonlinear hybrid learning environment is changing its structural parts and particularly:

(a) The relation between teacher and student, by establishing the principle of changing roles, either in the conventional classroom or the virtual one or its hybrid form.
(b) The relation of the teachers with teaching itself, since from a unique source of knowledge they evolve into coordinators of numerous sources of information, changing the learning objective from accessing information to the handling and processing of timely and reliable flow of information.
(c) The relation of the students with the new environment, where the two factors equation of course attendance - examinations is replaced by the three factors equation of active participation in the training procedure, creation of a personal hybrid learning area, multiple as well as two-way approaches of evaluation

Recent years have seen a tremendous increase in university distance learning courses taught over the Internet (Clark 1999). During the last thirty years, there has been quite some discussion on distance learning models, as distinct and separate from the traditional learning model. Some of these models include the methodology of independent study (Wedmeyer, 1977), the learner autonomy (Moore, 1977) and interaction and communication (Holmberg, 1989). A comparative analysis of all such theories recognizes their importance but realizes that they cannot be considered a complete pedagogic model (Keegan, 1993).

Nowadays, distance learning is an official educational system wherein students in different geographic locations communicate with their instructor and with each other through interactive media and telecommunication systems (Simonson et al, 2000). Computer-mediated communication takes several forms including electronic mail, electronic conferencing, video conferencing and the World Wide Web which can also be used for the submission and publication of students' assignments and of tutors' comments on students' work (Dede, 1996; Anastasiades, 2002c).

Table 1. Education Models- A Conceptual Framework (Anastasiades, 2003a)

	Teacher Participation	Student Participation	Face to Face Communication	Computer Mediated Communication
Face to Face Instruction				
Teaching – Lectures	YES	No	YES	NO
Monitoring - Questions	No	YES	YES	NO
Learning Material Development	YES	No	Printed	NO
Reading List Development	YES	No	Printed	NO
Web Based Education				
Teaching – Lectures	YES	No	NO	YES
Monitoring - Questions	No	YES	NO	YES
Learning Material Development	YES	No	NO	Digital
Reading List Development	YES	No	NO	Digital - Links
Hybrid Education				
Teaching – Lectures	YES	No	YES	YES
Monitoring - Questions	No	YES	YES	YES
Learning Material Development	YES	No	Printed	Digital
Reading List Development	YES	No	Printed	Digital - Links
Non Linear Hybrid Educational Model				
Teaching – Lectures	YES	YES	YES	YES
Monitoring - Questions	YES	YES	YES	YES
Learning Material Development	YES	YES	Printed	Digital
Reading List Development	YES	YES	Printed	Digital - Links

Interactive Videoconferencing (IVC) nowadays has a significant role in order to support interactive synchronous learning activities by distance. The Interative Videoconferencing process allows learners and instructors to share ideas and information, to cooperate, to collaborate at the same time from distant locations (Anastasiades, 2003c; Thompson, 1995; Martin, 2000). This means that emphasis is placed on the teacher as a cultural intermediary during the learning process in its new form, in the mean time the social consequences of the new technologies need to be constantly assessed (Anastasiades, 2003d).

This chapter aims to present the theoretical background and the holistic methodology which have been developed in the Department of Education at the University of Crete, to implement step by step a pedagogical model for IVC instruction during the last seven years.

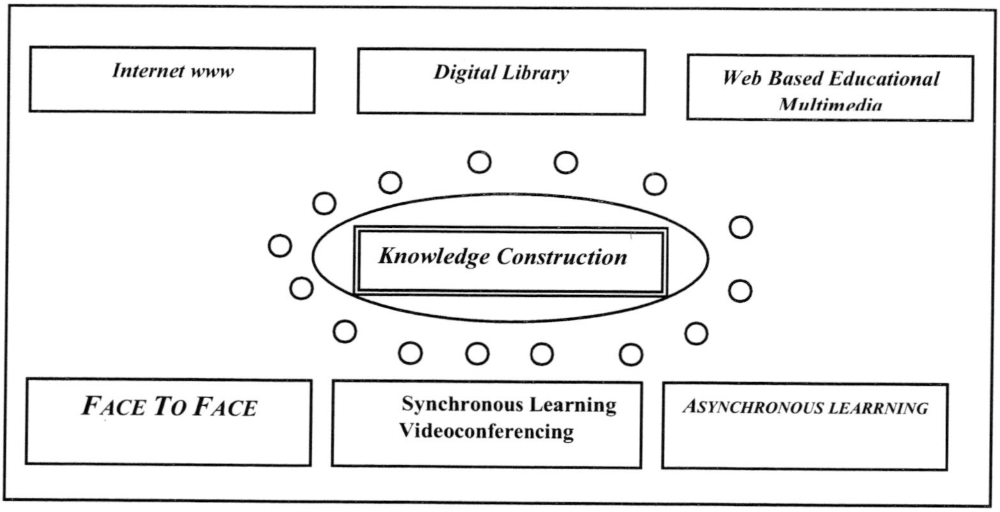

Figure 1. The New Learning Environment – a constructivism approach.

This chapter is organized as follows: in the second part we discuss the impact of ICT in Distance Learning process whilst in the third part will expound the content of the educational use of Interactive Videoconferencing (IVC) in Learning and Teaching. The fifth section will cover the theoretical framework of the proposed IVC pedagogical model The fourth section will cover the aspect of IVC instructional design and finally in sixth section we will present two IVC projects which take place in the University of Crete during the last seven years

2. ICT and Distance Learning: A Literature Review

According to Bates, (1995, p. 12). "Good teaching may overcome a poor choice in technology, but technology will never save bad teaching; usually it makes it worse. Good teaching matters."

The rapid development of ICT applications (teleconferencing, electronic file transfers, and more), the electronic interconnection of learning units and the penetration of web educational multimedia into the hard core of the learning process, is establishing the pedagogical challenges for the World Wide Web ((Lian, 2000, Hammond, 2000)) as one of the most important factors of major and radical changes in the learning environment in which advanced learning technologies meet pedagogy and collaborative learning (Bernard, et al, 2000).

New Technologies have contributed to the development of a third wave of distance learning obtaining new possibilities for the synchronous and asynchronous communication (Anderson, 2004; Berge and Collis, 1995α). This resulted to the broadening of the ways and methods of communication between learners and educators as well as ensuring the effectiveness of the educational tools (Anastasiades, 2005; Paulsen, 1998).

The way the technology is used in the distance learning has evolved a few questions due to the emphasis given the available technological tools and less to the educational aspects that should characterize such a learning environment (Ally, 2004; Kozma 2001, Anastasiades, 2003e)

The development of distance learning mostly in USA but also in Europe has been chronically equated with the rapid introduction of ICT in the educational world (Bates, 1995).

The dominant techno centric theories especially in the USA where teleconferencing is mostly viewed as a technological aspect and not as an organized entity does not keep pace with the general pedagogical culture of Europe about distance learning (Peters, 1998: 144; Bates, 1995:167).

The techno centric approaches have resulted to the development of educational environments which does not favor critical thinking resulting to the transmit of non critical social stereotypes with serious consequences (Massialas, 1989, Paulsen, 2003).

The emphasis given nowadays in the technological appliances could never replace the pedagogical dimension of teaching and learning.

New Technologies should don not only operate as a neutral supporting educational tool but also should be included in a general cultural and social framework (Lionarakis, 2006, Raptis & Rapti, 2004, Makrakis, 2000)

In conclusion, someone could claim that the dynamic and creative corporation of new technologies in a general social framework (Carr & Kemmis, 2002) and the use of technology under pedagogical circumstances could encourage the developing of interactive learning environments by distance which would provide learners with the ability to use their mental schema creatively and critically (De Bra et al, 2000; Brusilovsky, 1999; Kostoula & Makrakis, 2006), and to develop their own conditions so that students will become applicable to all forms of distance learning (Lionarakis, 1998).

3. Interactive Videoconferencing (IVC) as a Crucial Factor in Distance Education

3.1. IVC Definition

Although it is difficult to define the term videoconferencing (Coventry, 2000), there are many definitions for videoconferencing in the international literature:

- A videoconference (also known as a video teleconference) is a meeting among persons where both telephony and closed circuit television technologies are utilized simultaneously. Video teleconference communication is multi-way and synchronous, as it would be if all parties were in the same room. (en.wikipedia.org/wiki/Videoconferencing)
- Videoconferencing linking two or more locations by two-way video, allowing all participants to see each other simultaneously. (www.amideast.org/publications/aq/Back_Issues/1995/Sp95-html/Sp95Glossary.htm)
- A videoconference is a live connection between people in separate locations for the purpose of communication, usually involving audio and often text as well as video. At its simplest, videoconferencing provides transmission of static images and text between two locations. At its most sophisticated, it provides transmission of full-

motion video images and high-quality audio between multiple locations. www.tufts.edu/orgs/edmedia/gloss.shtml

Interactive Videoconferencing (IVC) technology allows students at two or more distant locations to create a collaborative environment at the same time (Suthers, 2001; Gibson και Cohen, 2003; Gowan, & Downs,1994; Reed & Woodruff, 1995, Alexander et al., 1999; Chandler and Hanrahan, 2000). The communication may include data and graphics exchange (Brown, 2001; Finn, Sellen, & Wilbur 1997; Andrews, T. & Klease, G. 1998) and data sharing (Gürer, et all 1999).

3.2. IVC in Learning and Teaching by Distance

Although it is difficult to define the term videoconferencing (Coventry, 2000), there are many definitions for videoconferencing in the international literature:

- A videoconference (also known as a video teleconference) is a meeting among persons where both telephony and closed circuit television technologies are utilized simultaneously. Video teleconference communication is multi-way and synchronous, as it would be if all parties were in the same room. (en.wikipedia.org/wiki/Videoconferencing)
- Videoconferencing linking two or more locations by two-way video, allowing all participants to see each other simultaneously. (www.amideast.org/publications/aq/Back_Issues/1995/Sp95-html/Sp95Glossary.htm)
- A videoconference is a live connection between people in separate locations for the purpose of communication, usually involving audio and often text as well as video. At its simplest, videoconferencing provides transmission of static images and text between two locations. At its most sophisticated, it provides transmission of full-motion video images and high-quality audio between multiple locations. www.tufts.edu/orgs/edmedia/gloss.shtml

Interactive Videoconferencing (IVC) technology allows students at two or more distant locations to create a collaborative environment at the same time (Suthers, 2001; Gibson και Cohen, 2003; Gowan, & Downs,1994; Reed & Woodruff, 1995, Alexander et al., 1999; Chandler and Hanrahan, 2000). The communication may include data and graphics exchange (Brown, 2001; Finn, Sellen, & Wilbur 1997; Andrews, T. & Klease, G. 1998) and data sharing (Gürer, et all 1999).

The result is that the participants have the right accumulated experience in order to use teleconferences in a proper and educational way especially in the university (Mitchell et. al 1993; Reed and Woodruff, 1995; Pitcher et all, 2000; Coventry,1998), for broadband matters (Smyth, 2005; Hearnshaw, 2000), and other related to financial matters (Twigg, 2002)

In universities, teleconferencing is soon going to be one of the most important ways of alternative teaching in the immediate future (Arnold et all, 2002; Heath & Halznagel, 2002; Marilyn et all 2002;

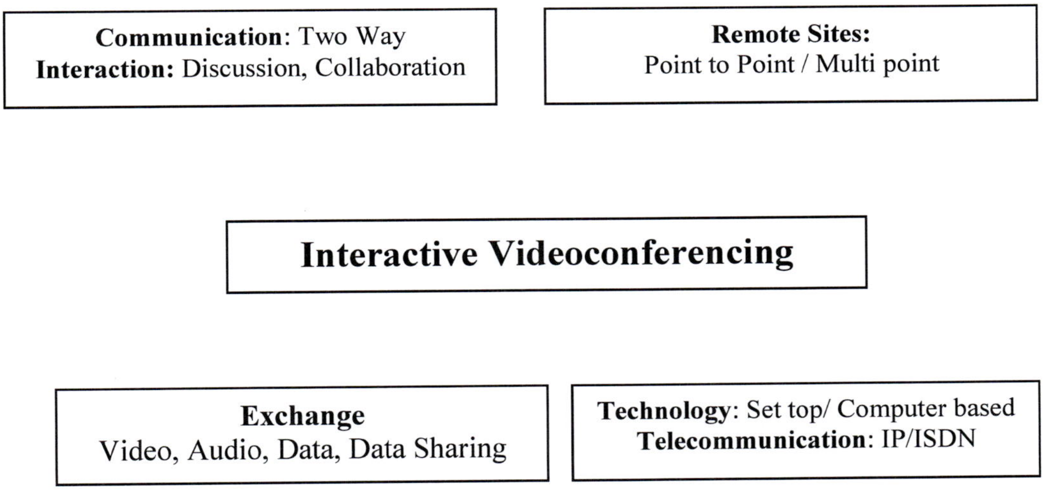

Figure 2. IVC Basic Characteristics.

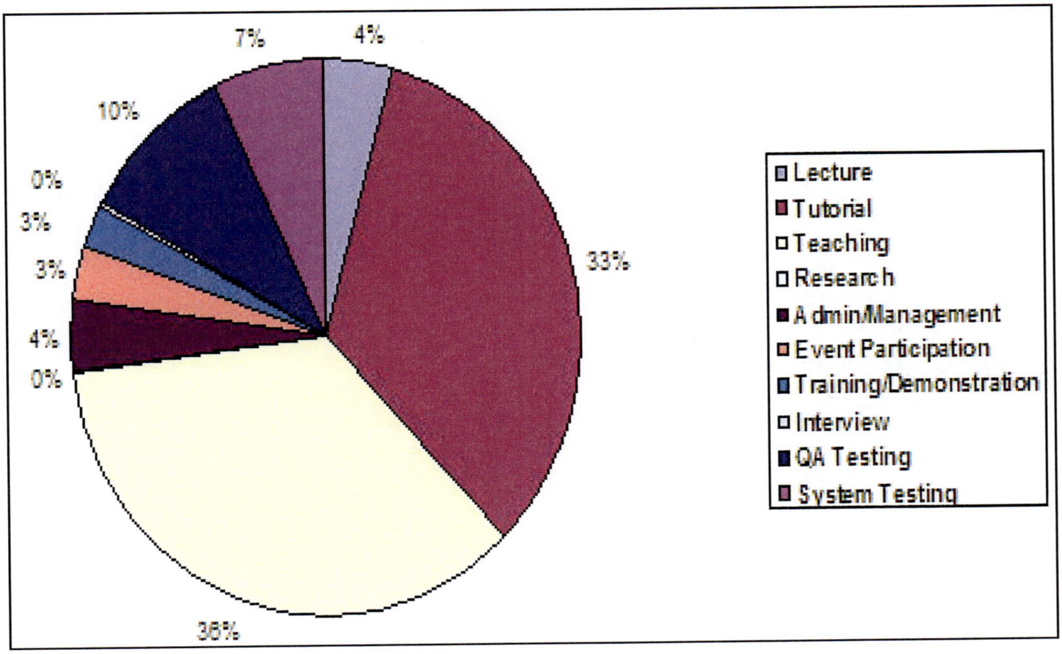

Figure 3. School Videoconferences via JVCS by Purpose (December 2006) http://www.ja.net/community/schools/vcsp/statistics.html

In 2006, according to a project conducted by Greenberg (2006) for the schools in the United States announced that 25% of elementary and secondary schools are having collaborative learning activities in Distance via Videoconferencing almost every day. Also, in England a survey conducted for the net JANET found that videoconferencing is mainly used for pedagogical and instructional purposes.

Table 2. IVC an Education: The three introductory waves of teleconferencing in the educational world

	Significant advances of technology	Utilization	Communication Teleconferencing System
First Wave 1990-1994	*1990 Dec*: CCITT standard H.320 for ISDN conferencing *1991*: IBM and PictureTel demonstrate videophone on PC *1991 Sep*: First audio/video conference (H.261 hardware codec) at DARTnet *1992 Sep*: CU-SeeMe v0.19 for Macintosh (without audio), by Tim Dorcey from Cornell *1993 Apr*: CU-SeeMe v0.40 for Macintosh (with multipoint conferencing) *1994 Feb*: CU-SeeMe v0.70b1 for Macintosh (with audio)	• conferences' attendance • Participation in administration meetings	ISDN Rooms T/Δ Set top
Second Wave 1995-1999	*1995 Aug*: CU-SeeMe v0.66b1 for Windows (with audio) *1996 Dec*: Microsoft NetMeeting v2.0b2 (with video) *1997 Jul*: Virtual Room Videoconferencing System (VRVS), Caltech-CERN project *1998 Jan*: ITU-T standard H.263 v2 (H.263+) video coding *1998 Apr*: CU-SeeMe v1.0 for Windows and Macintosh (with color video), from Cornell *1999 Apr*: Microsoft NetMeeting v3.0b (with gatekeeper)	• Participation in virtual dialogical classes	ISDN/IP Set top/ Computer based (Hardware Computer based (Software
Third Wave 2000-......	*2000 Dec*: Microsoft NetMeeting v3.01 service pack 2 (4.4.3396)) *2001 Sep*: World's first trans-atlantic tele gallbladder surgery (operation Lindbergh) *2001 Oct*: Microsoft NetMeeting v3.01 (4.4.3400) on XP *2001 Dec*: JVT video coding (H.26L and MPEG-4 Part 10) project, by ITU-T SG16/Q.6 (VCEG) and ISO/IEC JTC1/SC29/WG 11 (MPEG) *2002 Jun*: World's first 3G video cell phone roaming *2002 Dec*: JVT completes the technical work leading to ITU-T H.264 *2003 May*: ITU-T recommendation H.264 advanced video coding	• Interactive learning environments • Combination of synchronous and asynchronous learning environments	Platforms of synchronous education Synchronous interactive learning environments

Sources: Wall Street Journal (27 February 1996), The MBone FAQ, rem-conf listserv, The MBone listserv, CU-SeeMe listserv, RTP: Historical Notes, and few PostScripts (*.ps).

3.3. The Particular Features of the Teleconference's Educational Use

New Technologies support effective and educational uses of teleconferencing facilities by providing innovative learning strategies that would improve communication and interaction between the learner and the educator (Latchem, 2002; Kerrey & Isakson, 2001).

If someone believes that teleconferencing is like a conventional "face to face teaching" they will be disappointed. Although teleconferencing provides participants with the ability to watch, hear and communicate with each other at the same time, the interaction among the persons is more impersonal than in a conventional "face to face" teaching process (Bonk et al, 1998, Collins, 1991, Schweizer *et al.*, 2003). Participants do not share the same three-dimensional space, as they can watch only what the camera shows and the non verbal contact among them is usually vague (Bruce, 1996).

Another significant factor that creates more difficulties is that the participants can not evaluate their visual contact as when they see each others' eyes but not under physical conditions (Angiolillo et all, 1997).

The first studies that have been conducted by evaluating the effectiveness of the educational use of teleconferencing have found that it has not yet met the participants expectations (Motamedi, 2001).

Other studies claim that during teleconferences the participants of the distant classroom were not as satisfied with the learning process as the learners in the local classroom with the main technological appliances of teleconferencing (Knipe & Lee, 2002).

Also, according to some reports the conventional "face to face" teaching is more satisfying for some learners than teleconferencing (Delaney et all 2004).

From a technological aspect the audio-visual quality affects significantly the effectiveness of teleconferencing (Khan and Hirata, 2001). The last years, the development of teleconferencing tools, the continuing progress of audio-visual appliances and the revolution of broadband nets improve significantly the technological background when organizing educational teleconferences.

From the pedagogical aspect, teleconferencing shapes a new learning and teaching environment which is quite different from the conventional face to face teaching (Hearnshaw, 1999).

The complicated organizing of teleconferencing is a serious issue as many factors should be considered as the planning and organizing of the teleconference, human communication and the specialized education in relation to the continues development of technology and the high cost of such technological tools (Fincher, 2000).

We focus on the pedagogical framework which should be combined with modern technological tools and new methods and techniques in order to enhance interactive learning, to motivate learners to exercise control over their own learning experience and to become accountable for their own learning outcomes (Reeves, 1997)

4. Towards an IVC Constructivism Pedagogical Model

4.1. Why Do We Need a Pedagogical Model for the Educational Facilitation of Videoconferencing?

According to the above, the technology provided during a teleconference changes a lot the way of teaching and affects significantly serious issues of constructivism and organizing that educators follow in classroom, while the last in cooperation with the learners adopt new roles and become more skillful. Someone would conclude that the organizing of an educational teleconference is more demanding than the conventional face to face teaching (Health, 1999).

Berge και Mrozowski (2001), who studies educational teleconferences from 1990 since 1999 concluded that the most important issue for the successful outcome of a teleconference was not only the technology provided but also the educational methodology that was constructed and followed by the educators.

The main objective of videoconferencing is not to replace face to face conventional teaching but to come into supplementary operation (Anastasiades, 2006; Hanor & Hayden, 2003)

Many studies propose new ways of teaching in order to improve interaction among people while new advisory guides have been published in order to help participants to improve videoconferencing programs (Robinson, 1997; Hayden, 1999; Digital Bridges: K-12 videoconferencing)

This has resulted to new forms of learning that encourages learners and educators to work more interactively, creatively and actively (Clear, 1999, Ackermann, 1996). This also motivates learning through the development of interactive workgroups (Merrill et al, 1996; Andrews & Klease, 1998; Fischer *et al.*, 2000; Fischer & Mandl, 2003), between local and distant classes.

Recent research investigates how to foster collaborative learning through videoconferencing (Ertl ta all, 2005). Results indicate that content schemes as well as collaboration scripts were beneficial to collaboration outcomes.

More and more studies focus on the design of educational models adjusted to the needs of the effective and educational uses of teleconferencing.

According to Chen et all (2005), five dimensions of the model for synchronous learning can be identified. "These are: *role* (teacher, student), *participant* (individual, group), *venue* (specific, any), *interaction* (one-way, two-way), and *delivery* (playback, live)."

The proposed online synchronous live instruction task describes three types of interaction (Chen et all, 2005):

(a) Students ask questions and teacher responds
(b) Teachers assign questions and students respond
(c) Students Discuss online and interact with each other

Untill now there have been conducted many studies that focus on the particular features of teleconferencing and suggest new ways and methods effective and educational uses of teleconferencing facilities. Though, there is a serious lack of researches (lack of pedagogical research) that study the instruction and implementation of an holistic Pedagogical Model for the development of effective and educational uses of teleconferencing in the educational environment

4.2. The Theoretical-Philosophical Framework of the Suggested Pedagogical Model IVC

The suggested Pedagogical Model is based on the principles of cross curricular thematic approach and on the constructivism approach. The promotion of cross curricular thematic approach through the contemporary educational theories is the result of the transition from the philosophical frame of determinism, cartecius perception of things and the technological determinism in recent times to the new reality of the inter-disciplinary approach, of Gestalt psychology and the social transformation (Matsagouras, 2002).

According to the cross curricular thematic approach, the scientific knowledge of various subjects and problematic situations, is one's attempt to get to know with themselves better as well as the reality that surrounds them aiming at their personal development and social corporation through gaining knowledge, attitudes and beliefs. (Beane, 1997)

The materialization of cross curricular thematic approaches should be accompanied by the design of a linear developmental model of a teaching system at the center of which there will be empirical, holistic and collective approaches.

According to this approach, the exploitation of Constructivism Theories, Situated Learning and Learning Communities is proposed.

Constructivism Theory can serve cross curricular thematic schemas of teaching approaches for two main reasons (Matsagouras, 2002: 37):

(a) According to the principles of constructivism, learning should occur in authentic situations and to be in agreement with the personal interests of students and teachers, a practice that is in favor of the Cross Curricular Thematic Approach.
(b) Constructivism approach considers that knowledge is constructed due to integration, which is the incorporation of new information in the already existing cognitive schemas according to which learning is promoted through the relations of the parts and the holistic approaches of the units.

The Theory of Situated Learning supports that learning is implemented through the learners' participation in "communities of practice" or "learning communities" and is the result of the combining of activity, environment-frame and culture where is taking place (Lave, 1997; Lave & Chaiklin, 1993; Wenger, 1999 and Wenger & Snyder 2000).

The development of social and cognitive skills can be promoted through the approach of cognitive apprenticeship (Brown, Collins and Duguid 1989), at the centre of which there is collaborative social interaction, connection of knowledge and action and social structure of knowledge through the learner's involvement in authentic situations.

The connection of our teaching approaches with the students' interests and planned collective action in the spectrum of cross curricular thematic approaches can be supported by the method of project (Matsagouras, 2002; Kanakis, 2001) which will be analyzed in unit 5.

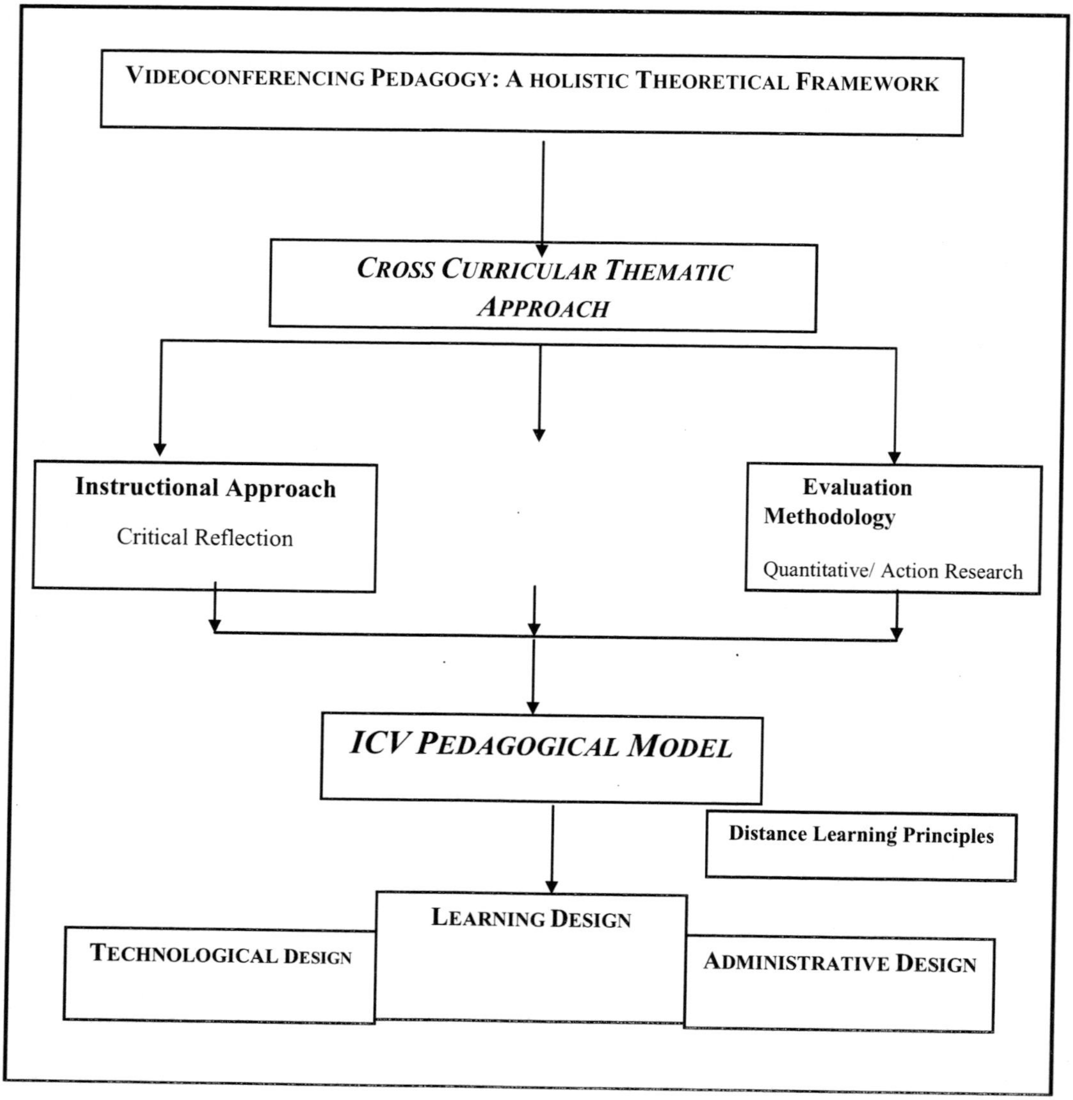

Figure 4. Videoconferencing Pedagogy: A Holistic Theoretical Framework.

The propose methodology helps a quality of learning tools and methods to be developed according to the teaching goals and the nature of the teaching concept.

4.3. Main Constructive Principles and Phases of Development of the Pedagogical Model IVC

The conventional constructive teaching models are linear or structured and were proposed in order to be content with the traditional principles of objectivism and positivism according to Dick & Carey (1996) and Tyler (1949).

Table 3. Developing Phases of the Pedagogial Model IVC

1st Phase: Needs Analysis
Action 1: Investigation of the classes´ Needs for the Participation in the Teleconferencing program
Action 2: Investigation of the Teaching Needs
Action 3: Analyzing the subordinate technological characteristics of the school
Action 4: Investigation of the elements that characterize the culture and attitudes of the school´s/classes environments
Action 5: Investigation of the teachers´ characteristics who will participate in the teleconferences
Action 6: Analysis of the subordinate characteristics of the teachers and learners of the target group
Action 7: Analysis of the subordinate Administrative and Organizing Parameters
2nd Phase: Pedagogical Design of IVC
Phase 2A: Designing the Teaching Methodology
Stage 1: Planning the Preparation Period
Action 1: Finding the school class (distant partners)
Action 2: Teachers´ Training
Action 3: Investigation of a General Thematic Unit
Action 4: Planning the Teaching Stages (Teleconferencing Pyramid)
Action 5: Definition of the Communication Model
Action 6: Architectural Design of the Class
Action 7: Preparing the students and their parents
Stage 2: Organizing the Teleconferences
Action 1: Definition of the Time-schedule
Action 2: Teaching Process
Action 3: Organizing the Educational Material
Action 4: Planning the interaction activities
Action 5 Formulation of Evaluation
Stage 3: Organizing an holistic Evaluation
Phase 2B: Technological Design
Action 1: Specifications of the teleconference/system´s characteristics
Action 2: Specifications of the Communication Infrastructures
Action 3: Specifications of the demanding additional technological equipment (screens, projectors etch.)
Action 4: Specifications of the demanding software
Action 5: Formulation of Evaluation
Phase 2C: Financial Planning
Action 1: Drawing up the Budget – Time-schedule
Action 2: Finalization of the procedure for the supplies, purchase, installation, service
Action 3: Selection of the most convenient solutions
Action 4: Formation of Evaluation
Phase 2D: Adminisrtative – Organizing Structure
Action 1: Instruction of an organizing schema that will govern the participants´ relations
Action 2: Formation of an observation board
Action 3: Formation of an educational organizing board
Action 4: Formulation of Evaluation
3rd Phase: Implementation of Teleconferences
4th Phase: Evaluation Sumative Evaluation - Revision
In the next unit we will focus on the 2nd Phase of the pedagogical design of educational teleconferences.

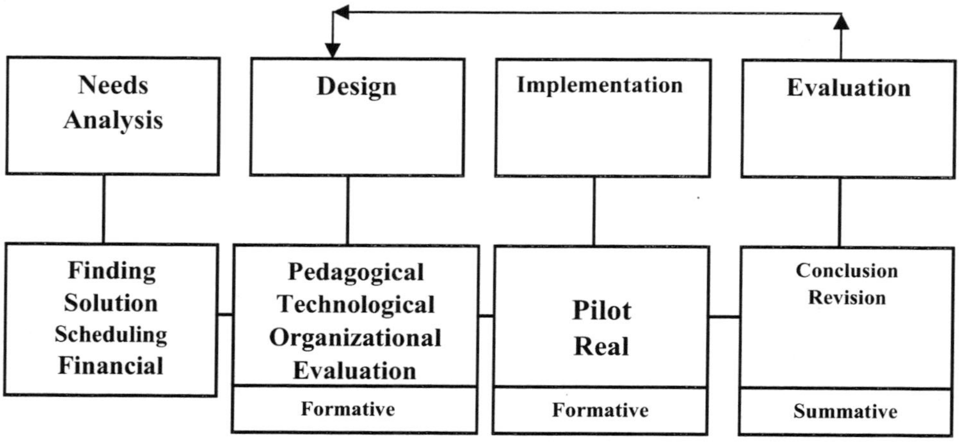

Figure 5. The Non Linear IVC Pedagogical Model.

According to the analysis presented in unit 4.2 our main goal is to plan an open learner-centered teaching system of constructivism. The teaching plan of this system should be structured on the basis of linear models which are very popular to use these days and are based on the theory of constructivism (Piaget, 1970; Blumer, 1969; Kuhn, 1996; von Glasersfeld, 1989; Vygotsky, 1978).

According to Vrasidas et all (2000), the basic characteristics of this model are:

1. The design of the teaching system is based on particular phases and actions
2. Specific administrative methods of the development work program, usually adopted by the development of information systems
3. The training systems are consisted of the following three parts:
   (a) human funds,
   (b) funds for instructive materials, and
   (c) funds for technological infrastructure
4. The four phases of training plan are the analysis, formation, materialization and evaluation.

On the basis of our philosophical and educational theoretical framework and especially for the particular pedagogical and technological demands of the educational use of teleconferencing, the following model of a developing training system was designed (Vrasidas & McIsaac, 2000).

The developing phases and suggested actions followed are described in Table 3.

5. IVC Pedagogical Design- Basic Characteristics

According to the previous analysis in unit 4.2, a constructivism model IVC is suggested which is based on the following linchpins:

1. Framework of Distance Learning Principles (ADEC, 1999)

The proposed methodology adopts the ADEC Guiding Principles for Distance Teaching and Learning and the current methodology accepts the following principles:

- The learning experience must have a clear purpose with tightly focused outcomes and objectives. Web-based learning designs must consider the nature of the content, specific context, desired learning results and particular characteristics of the learner. Learner-centered strategies include modular, stand-alone units that are compatible with short bursts of learning. Learning modules may also be open, flexible and self-directing.
- The learner is actively engaged. Active, hands-on, concrete experiences are highly effective. Learning by doing, analogy and assimilation are increasingly important pedagogical forms. Where possible, learning outcomes should relate to real-life experiences through simulation and application.
- The learning environment makes appropriate use of a variety of media. Various learning styles are best engaged by using a variety of media to achieve learning outcomes. Selection of media may also depend on the nature of content, learning goals, access to technology, and the local learning environment. .

2. Cross Curricular Thematic Approach

On the basis of our theoretical framework we chose the Cross Curricular Thematic Approach (Matasagouras, 2002)

3. The Social construction of Knowledge

Learning as a cognitive procedure develops in a social environment. On the basis of the constructivism approach, teaching should be designed according to the actions and experiences that the learner undergoes with the immediate learning and broader interactive social environments.

4. The Project Method

The Project Method encourages the development of dynamic environments which support cooperative learning through critical thinking, (discovery learning), and the learner's involvement in authentic situations enhancing the student's development of personality, self-initiated action and critical thinking on the basis of an holistic approach of knowledge (Maxim, 1999; Knoll, 1997)

5. Communities of Practice?

Communities of Practice complete the goals of the Project Method as they encourage the construction of knowledge within the community in relation to the equal participation of persons and the creative exchange of views (Scardamalia, 2002) through a commitment procedure of practice community (Wenger,1998

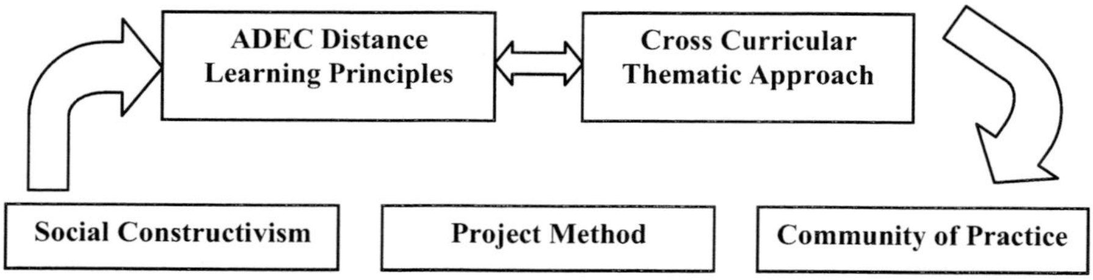

Figure 6. The Basic Characteristics of the Proposed Pedagogical Approach.

6. THE PROPOSED MODEL OF INSTRUCTIONAL DESIGN

The teaching structure of teleconferencing according to the Project Method should be based on the following related factors, such as:

- How do we define the general and specific objectives of a teaching process according to the main principles we have already referred?
- Under which criteria do we choose our subject?
- Taking on account that there are one or more distant audiences, which are the basic characteristics of our communication model that we will choose?
  Which is the relation between teachers and learners in a local or distant audience?
  Which is the relation between a local and a distant audience?
  How can we replace the physical absence of teachers and learners especially for educational and designing purposes?
- How shall we plan our teaching in order to achieve our main teaching goals?
  How shall we introduce participants to each other in order to help them interact sentimentally?
  How shall we design our teaching in order to make teachers and learners to interact as much as possible and particularly learners too?
  In what ways shall we organize collaborative activities and involve learners in authentic situations?
- The construction of a teleconferencing classroom can play a significant role in the achievement of our goals. Which is the most appropriate architectural design of our visual classroom?
- Which is the most proper methodology for our educational and teaching aims?
- What kind of interactive activities shall we choose?

7. PREPARING AND PLANNING AN IVC

The main goal of the proposed steps is to develop a teaching model adjusted to the needs of the educational IVC. The proposed structure of educational teleconferencing is the result of a 7-year old teaching experience, that took place in 30 elementary schools, while the necessary theoretical documentation adopts critical approaches from the cross curricular thematic approach and the Project method of Matsagoura (2002) and Frey (1998).

7.1. Action 1: Finding a Class (Distant Partners)

It is not easy to look for classes in order to design an interactive teleconferencing between teachers and students, as the criteria proposed in unit 4.3 Phase A (Analysis of opportunities and needs) should be carefully followed.

Searching for classes can be accomplished either through announcements in relative forums (ED1VIDCONF Digest), or through special designed data bases which include various criteria of searching as the language, existing technological appliances, projects etch. (www.etwining.gr).

In these cases the results are of questionable quality as the educators are not usually properly trained or skilled.

The existence of an educational organization that would provide the right partners and would support educators and learners in technical issues would be of great importance for the accomplishment of such endeavors.

The last years, the Department of Education of the University of Crete has implemented the first program on distance learning in elementary schools, called "ODYSSEUS" and has supported many schools that took part in teleconferences to find partners and to provide them with educational and technical support.

The criteria for choosing schools that would participate in teleconferences have been already referred in unit 4.3 and include:

- Broadening the Necessity of the Classes' Participation in the Program of Teleconferencing
- Investigating Teaching Principles and analysing the school's technological appliances
- Investigating factors that characterize the culture and behavior of school's/ classroom's habitats
- Investigating the basic characteristics of teachers who will participate in the videoconference
- Analyzing the teachers' basic characteristics of the target group
- Analyzing the technological, administrative and management parameters

On the basis of the present methodology the collaboration of a pair of schools is proposed.

7.2. Action 2: Teachers' Training Program

The educational organization designs a plan for the distance training of educators who will participate in the teleconferences and aims to their familiarization with the new learning environment and to acquire all the necessary knowledge and skills in teaching and technical level.

After the end of the training program the educators should be in position to:

- Identify the concept of Cross Curricular Thematic Approach
- To design the project method through videoconferencing
- To identify the basic pedagogical and technical features which are necessary for the educational use of a teleconferencing system in their classroom
- To develop techniques and methodologies adapted to their specific teaching demands

7.3. Action 3: Investigating a Subject – General and Specific

Educators from distant classes communicate through videoconferencing, discuss in pairs about subjects of a general thematic context e.g. the sea, the environment, culture etch. According to the students' general social and cultural background, the subjects discussed are usually related to the students' existing knowledge and interests.

In the end, a conversation is taking place according to the general and specific aims of the teleconferences. The basic aim of this action is to prepare the students of the distant classes to make better conversations.

7.4. Action 4: Planning a Communication Model

On the basis of the chosen subject, when the first goals are set and the technological equipment of the interacting schools are provided, the choice of a communication model depends on the content of the teleconference.

The communication model defines the following:

1. The number of classrooms that will participate in the videoconferencing: The videoconferences should take place in a pair of classes for educational and practical reasons.
2. The educators' role in every classroom: The role of the educators is coordinating, advisory.
3. The students' role: Students are in the centre of the learning process, they plan and materialize their activities.
4. The kind of communication connection

   The connection through Internet (over IP- minimum download 256/ upload 1024) is cheaper for schools but unfortunately it has not a good quality of vision and sound.

   The ISDN (min 2ISDN lines) connections have a stable quality of vision and sound but they are quite expensive too.

   The combination of the above solutions has been proved to be quite effective.

   The kind of teleconferencing systems The teleconferencing systems should:

   Support the connections through Internet (over IP) as well as the ISDN connections

   Support the data transfer and the shared use of applications

This proposal is combined with the application of the three models method of Maryland University, USA, and particularly model A (IDE, 1996) that is a Virtual Classroom composed of groups of students in two distant locations.

7.5. Action 5: The Instructional Stages (The IVC Pyramid)

Teachers and students will get in contact with the new IVC constructional model progressively, in order to become a part of the new learning environment as smoothly as possible.

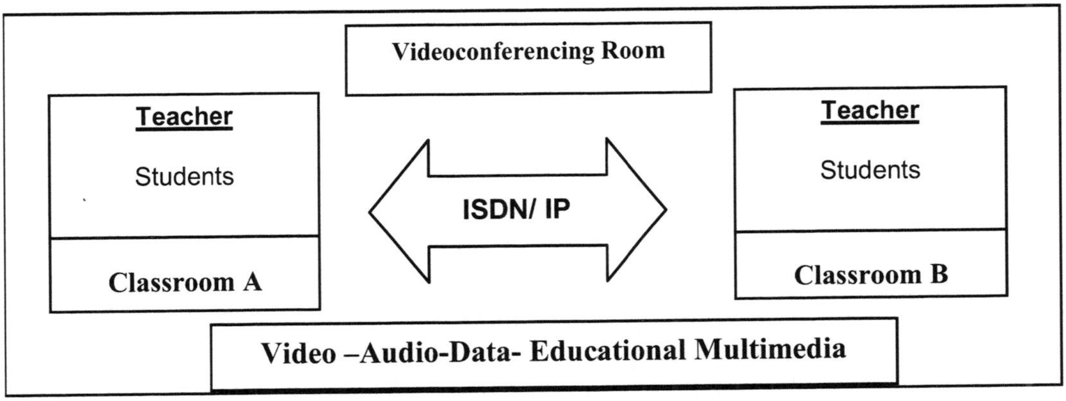

Figure 8. The Proposed Communication Model.

Picture 1. The Proposed Communication Model (From theory to Practice - ODYSSEAS 2004).

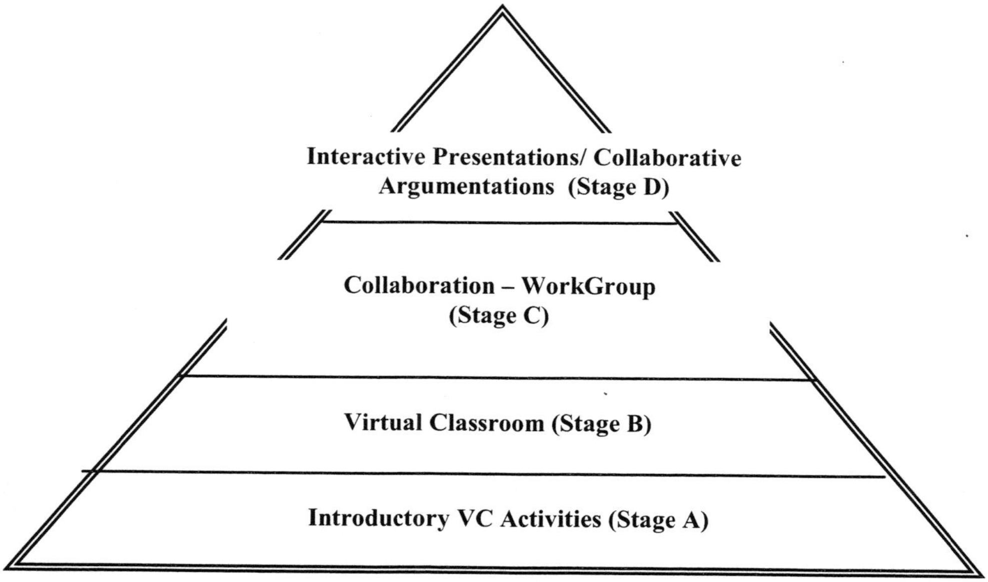

Figure 9. The IVC Pyramid.

The proposed Project Method will be achieved through the implementation of four Stages:

Stage A: Introductory VC Activities

In the first stage, language activities are designed for the development of language, kinesthetic and musical intelligence.

The main target of the first stage is:

- Educators and learners to become familiarized with the new learning and teaching environment.
- To be introduced to the students and to introduce the students to each other by sensing their social and cultural background.
- To begin an open conversation among students according to the method of Brainstorming so that students will focus on a subject of their interest and to be involved in authentic situations. We will analyze this subject in the following teleconferences.

Stage B: Virtual Classroom

In the second stage, the students in groups try a first cross curricular thematic approach of the subject dividing it into smaller units. The analysis of the subject can be based either on their interests or on the curricular program, or on the theory of multiple types of intelligence (Gardner (1993).

The investigation of the students' main ideas in relation to the proposed subject and their involvement in authentic learning environments are some of our most important priorities.

In this stage, the design of activities emphasizes the construction of knowledge which is based on the content as well as on the social frame according to which learning occurs (Content and Context dependent Knowledge) (Johansen, 1999).

The teaching plan in phase B focuses on the cross curricular thematic approach of knowledge, the students' involvement in authentic situations and of critical productive confrontations and reflection, while teachers provide the proper support and feedback on the basis of cognitive apprenticeship and Scaffolding (Solomonidou, 2006). A formative evaluation follows

Stage C: Collaboration- Workgroups

In the third stage, students should be divided in to workgroups defining the unit of their interest. The groups of the local and distant class work collaboratively in order to define their main themes of work.

After the end of the teleconference groups agree to leave some time to prepare before beginning teleconferencing again.

During this time in internal team level, the steps are designed, the roles and tasks of the members of the team. A searching plan is organized for gathering information, defining the appropriate procedures, tasks and roles of the members of the group. This will help students to "learn how to learn" by using the printed as well as the electronic sources.

Finally, a composition and production of the results for every group separately is attempted.

A formative evaluation follows.

Stage D: Interactive Presentations/ Argumentation

In the fourth stage a two-parameter videoconferencing is being planned.

1. Presentation of learning tasks

The groups of the local and distant class present their work to their schoolmates and guests (parents, teachers, students of other classes) and teachers encourage argumentation between virtual classrooms. The learning tasks presented during videoconferencing is a significant way of rewarding the students (Fincher, 2000).

2. Collaborative Argumentation

Learners have to explicate their position focusing their cognitive activities on the problem so that different perspectives are essential to discuss them collaboratively (Fischer et all, in press)

3. Evaluation

It is very important to evaluate the result produced as well as the procedure of its production and the students'general impressions of their participation in videoconferencing.

A Summative evaluation follows

The teachers of the two classes define the number of teleconferences which are necessary for the proposed stage in order to maintain their goals.

7.6. Action 6: The Classroom Architecture Model

The arrangement of the classroom is one of the most significant parameters for a successful videoconference. The proposed model is a better edition of a previous one (Anastasiades, 2003) because the teleconferencing classroom is in the school.

This means that:

- This ordinary class contains 20- 25 students which if it is properly arranged it can be used as a teleconferencing classroom.
- The existing equipment of desks and chairs is used.
- The teacher does not usually have the technical support and that for the teacher should manipulate the equipment through a video control.

According to the above restrictions the following is proposed:

1. The arrangement of the classroom should help the development of interactive activities and to satisfy the needs of every teleconferencing.
2. The arrangement according to the plan......is meant to divide the students in to kinds of audiences.

   In audience A, students participate in collaborative exercises with other students when asked.

   In audiences B (B1+B2), students of the local class participate in collaborative exercises with each other.

   It is worth mentioning that students work alternatively from class A to class B according to the tasks and roles agreed.
3. The camera should be better placed over the screen because it helps the teacher and students to participate actively during teleconferencing.

Picture 2. The classroom architecture model (From Theory to practice ODYSSEAS 2004).

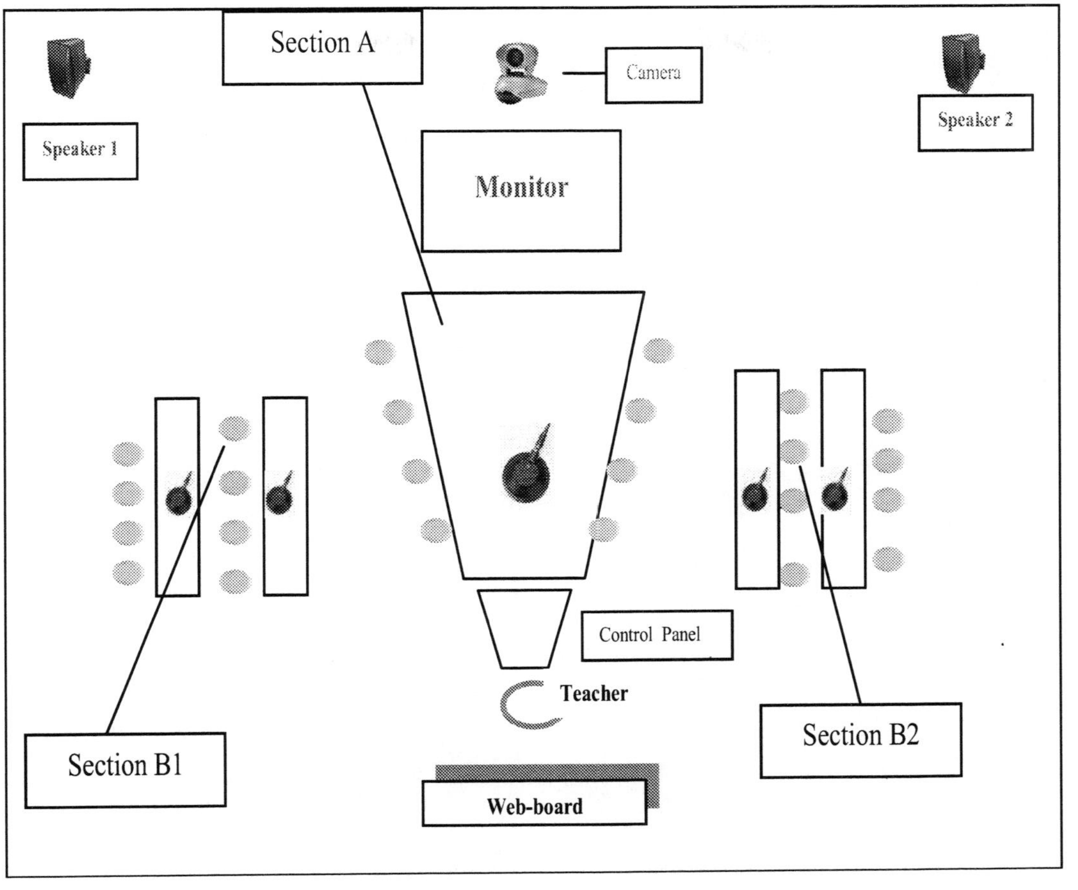

Figure 10. The classroom architecture model.

According to the proposed methodology four types of interaction are supported:

A. Interaction between students in the local classrooms
B. Interaction between students in the local and distant classroom
C. Interaction between students and teacher in the local classrooms
D. Interaction between students of the local classroom and teacher of the distant classroom

7.7. Preparing Students and Parents

Teachers have to discuss with students and parents about the new collaborative environment via IVC and try to prepare them with new reality.

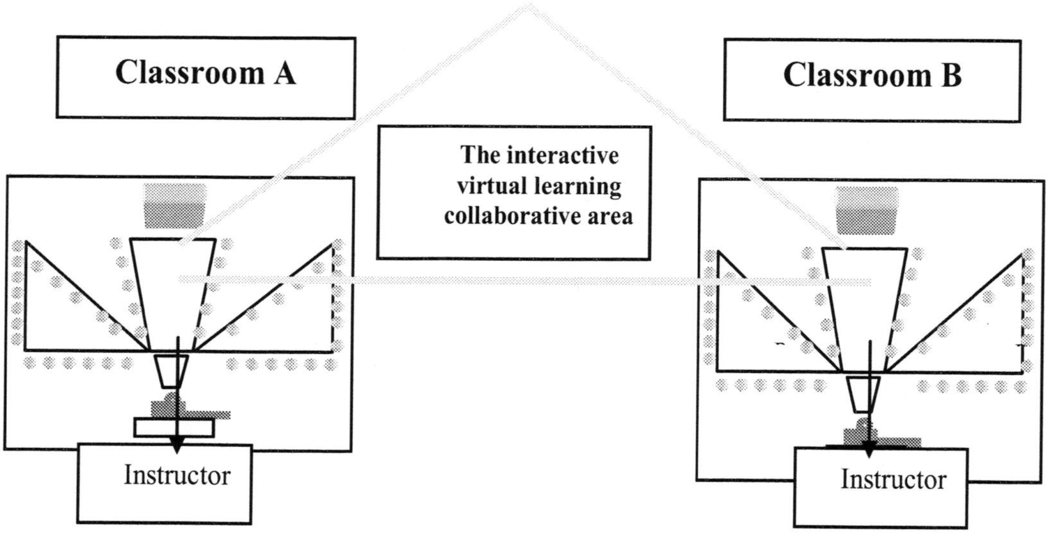

Figure11. The interactive virtual learning collaborative area.

Picture 3. The interactive virtual learning collaborative area (From Theory to Practice: ODYSSEAS 2004).

8. Evaluation Methodology

The elaboration of an evaluation methodology for IVC is one of its most important components. It was a decisive factor in revealing certain weak points in the teaching and pedagogic methodology at pilot stage, and at the same time it brought to light unknown, heretofore, instructional and social aspects of our endeavor.

Table 4. The 11 steps of "ODYSSEAS" evaluation methodology

Evaluation steps	Description
1. Evaluation target	Improvement of learning system methods and tools
2. Object of the evaluation (definition of the evaluation's scope).	Evaluation of the teaching and pedagogic approach Recording and evaluation of the potential social impact
3. Subject of the evaluation	Students

(who gets evaluated)	Teachers School Technological equipment and educational tools
4. Evaluation performers (who evaluate)	Students Teachers Support team Research team
5. Safety valves of the evaluation	Development of techniques to secure the reliability of the results
6. Evaluation criteria	Evaluation of students' knowledge and skills Evaluation of teachers' knowledge and skills Evaluation of technological equipment Evaluation of the teaching methodology by the students Evaluation of the teaching methodology by the teachers Evaluation of the social impact by the students Evaluation of the social impact by the teachers
7. Evaluation against time	Ex ante On Going Ex post
8. Evaluation tools	Questionnaires Reports Group interviews
9. Data collection – Data processing-analysis	Collection of data is immediate and continuous Spreadsheets Quality – quantity analysis
10. Presentation of results	Homogenized charts Structured reports Suggestions for improvement actions
11. Utilization of results	Immediate Ex post

The most widely used method for the evaluation of distance learning programs concerns higher education and introduces a four-point evaluation system (Keegan & Rumble, 1982). Evaluation focuses on the quantitative data of the educational institution that provides distance-learning programs, on the quality of the education provided, on the prestige of the said educational institution and, finally, on the relevant cost of the education provided.

Moreover, a considerable number of scientific projects have focused on the evaluation of distance learning programs (*Bobby Malone et al, 1998, F. Rowland, I. Rubbert, 2001, D. Henderson, 2001, Dasher-Alston et al, 1998).*

The evaluation methodology proposed as part of ODYSSEAS program follows a procedure of 11 steps(Anastasiades 2003b) which are the following.

However, the peculiarities of a distance-learning program implemented in an elementary school necessitate the development of a complete methodology that will answer the special needs and requirements of such an endeavor.

The evaluation methodology proposed as part of ODYSSEUS program follows a procedure of 11 steps, which are the following:

1. Evaluation Target

Main target of the evaluation is to detect elements and data that will contribute to a continuous improvement of the teaching methodology in real time.

2. Object of the Evaluation

The object of the evaluation is divided into two main levels: teaching-pedagogic and social.

Objective of the proposed methodology at a teaching level is the evaluation, on the part of the students and instructors, of the teaching methodology that was used. Pointing out the weak points in the planning and implementation of the various subjects to be taught, spotting good practices that should be promoted further and, finally, estimating the friendliness, efficiency and effectiveness of the teaching methodology were the key points of the evaluation.

At a social level, we were concerned about whether the distance learning process changed – and in what direction – the impression and idea of the students about their teacher, schoolmates and the area of their school as an institution of socialization.

3. Subject of the Evaluation

The evaluation methodology is targeting the students, the instructors and the components of the new learning environment (technological equipment, educational technologies, subjects etc.)

4. Evaluation Performers

Evaluation is carried out by the students themselves, the teachers, the support team (university

5. Safety Valves of the Evaluation

The introduction of the new communication and information technologies in elementary schools requires special attention. The evaluation procedure of a distance learning program becomes difficult, since the trainees, due to their age, cannot easily grasp its importance. In order to secure the targets of the evaluation, the teachers and the program manager need to apply considerable effort, both during the planning and the implementation stage.

So in order to secure the reliability of the results, four safety valves are being introduced. The first includes the special mention of the teachers to their students about the importance of the questionnaires. The second relates to the time allowed for the students in order to fill out the questionnaires after each teleconference session. The third safety valve provides for a very small number of pilot questions, whose answers are considered given, to a reasonable degree. Thus, considerable deviation from these answers suggests low reliability of the results. Finally, a limited number of questions repeat themselves with different wording. Purpose of this technique is to ensure that the questionnaires are filled out as carefully as possible.

6. Evaluation Criteria

The methodology worked out as part of ODYSSEUS program introduces the following evaluation criteria, whose purpose is to answer the priorities described above, that is the evaluation of the program both at a level of teaching methodology and at a level of social impact.

Evaluation of the students' technological statu

Ascertaining and recording the knowledge of the students participating in the program as to the new communication and information technologies.

Description: Through a set of questions, we seek to find out what do students know about computers, the possibility and usage level of the basic software applications, internet and other features, such as the use of educational software, participation in teleconference sessions etc.

Target: To record the students' technological knowledge and abilities. Method: Self-evaluation of the students through a questionnaire.

Implementation: Planning of easy and comprehensible questions with scales of answers: Excellent, Very good, Good, Fair, None.

Evaluation of the Teachers' Technological Status

Ascertaining and recording the knowledge and skills of the teachers taking part in the program.

Description: Graduate and postgraduate studies, special knowledge in educational technologies etc.

Target: To record the teachers' technological knowledge and skills. Method: Self-evaluation of the teachers through a questionnaire.

Implementation: Planning of questions with scales of answers: Excellent, Very good, Good, Fair, None or, alternatively, analysis of curriculum vitas.

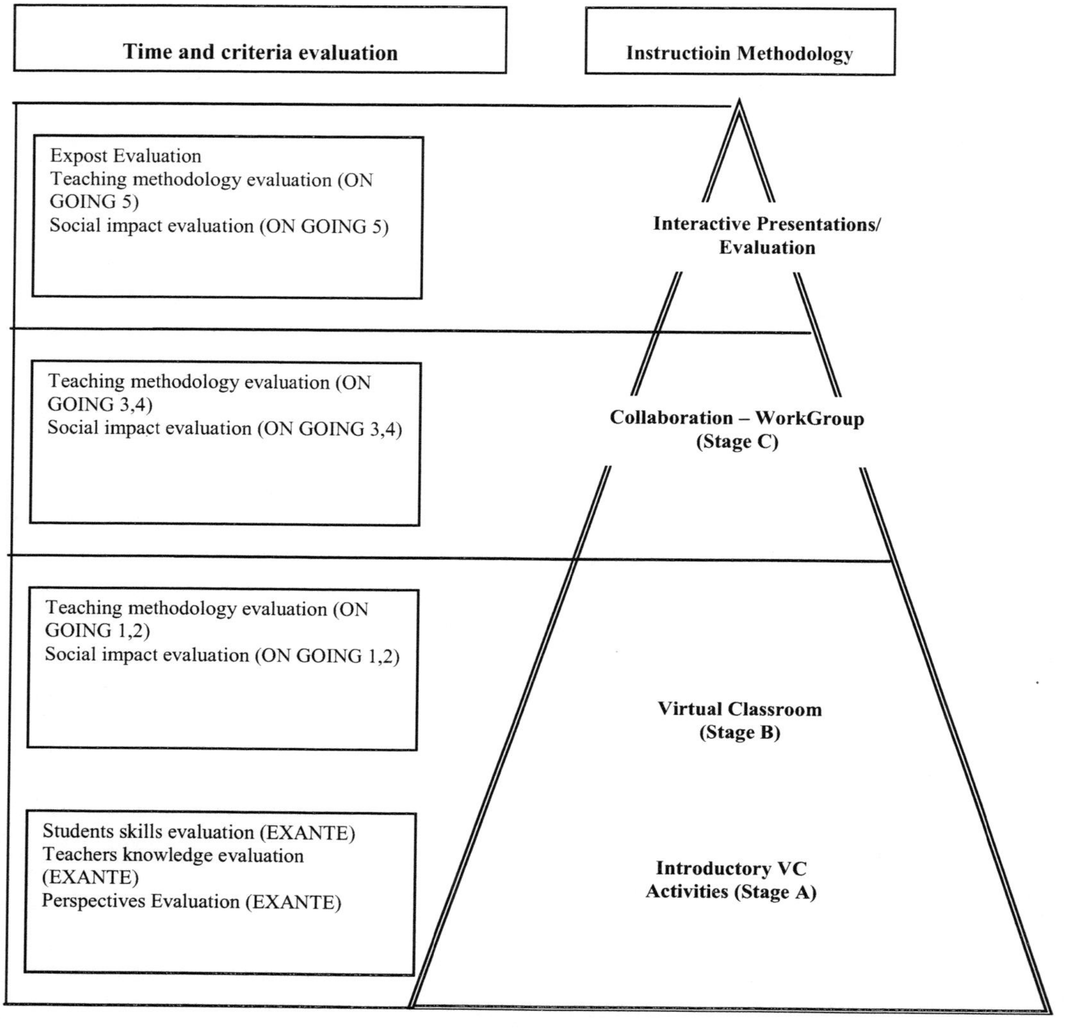

Figure 12. The relation between the teaching methodology, evaluation against time and evaluation criteria.

Evaluation of technological equipment

Recording potential technical problems.

Description: Detecting any technical problems in the evaluation system (sound, image etc.).

Target: To spot and remedy any technical problems.

Implementation: Development of reports.

Evaluation of the teaching methodology on the part of the students: A close look into the friendliness, usability and effectiveness of the teaching methodology.

Description: Through a set of questions, we seek to find out whether the students found it easy and pleasant to attend the course in its new form, what was their impression on the teachers' communication level, if they got distracted during the course of teaching etc.

Target: To spot any weak points and indicate good practices during the implementation stage of the teaching methodology.

Method One: Self-evaluation of the students. Implementation: Planning of easy and comprehensible questions with scales of answers: I totally agree (5), I agree (4), I don't know (3), I disagree (2), I totally disagree (1).

Method Two: Evaluation of students by their teachers. Implementation: Verbal questions and discussion between teachers and students during the lesson period.

Evaluation of the teaching methodology on the part of the teachers: A close look into the effectiveness and usability of the teaching methodology.

Description: We seek to record the teachers' views as to whether the new educational environment was friendly and pleasant or not, the attention level of the students, the evaluation of the courses' preparation etc.

Target: To locate any weak points and indicate good teaching practices during the implementation stage of the teaching methodology.

Method: Self-evaluation of teachers.

Implementation: Development of reports.

Evaluation of the Social Impact from the Students' Standpoint

A close look into the social impact and the relations between the students and their school environment in its new form.

Description: Based on a number of structured questions, there was an effort to outline the impression of the students regarding the schoolmates of their own class and also the schoolmates of the other class they contacted through teleconference, their teachers etc.

Target: To outline the impression of the students about their school as a place of incorporation into the society, before, during and after the end of the teleconference sessions.

Method One: Self-evaluation of students. Implementation: Planning of easy and comprehensible questions with scales of answers: I totally agree (5), I agree (4), I don't know (3), I disagree (2), I totally disagree (1).

Method Two: Evaluation of students by their teachers. Implementation: Verbal questions and discussion between teachers and students during the lesson period.

Evaluation of the social impact from the teacher's standpoint

A close look into the social impact and particularly into the relationship of the students with their teachers and the relationship of the teachers with each other.

Description: Recording any changes in the attitude of the students during the conventional courses, as well as during teleconference sessions, level of relations between teachers and students etc.

Target: To outline the teachers' idea of the students and their school as a place of incorporation into the society, before, during and after the end of the teleconference sessions.

Method: Self-evaluation of teachers.

Implementation: Development of reports

9. "ODYSSEAS 2004" IVC PROJECT: GENERAL DESCRIPTION

9.1. General Description

ODYSSEAS IVC project is the pivot of a collective endeavour, with the participation of the Ministry of Education of Cyprus the Department of Computer Science of the University of Cyprus, and the department of Education at the University of Crete. The main objective of this program is to form a cultural and intellectual bridge over schoolchildren from Cyprus, Greece and all around the world. (Anastasiades 2003a)

The students and teachers of ten elementary schools in Greece and Cyprus that participated in the project during 2000-2004 were the main component parts of an open collaborative environment by distance through videoconferencing. (Anastasiades et all, 2003c&2003d

The elementary schools that participated in the pilot program were Helenio Elementary School of Nicosia, Elementary School of Rethymnon, of Limassol, of Pissori, of Paphos, of Augorou and st Antonios,.

The main objective of ODYSEAS IVC project is to form a cultural and intellectual bridge over schoolchildren from Cyprus, Greece and all around the world. (Anastasiades, 2003).

9.2. Development of the Instructional Methodology – Implementation Process

The students will get in contact with the new teaching model progressively, in order to become a part of the new learning environment as smoothly as possible. This will be achieved through the implementation of four stages: a preparatory period, distance teaching, virtual class and tele-cooperation (Anastasiades, 2003).

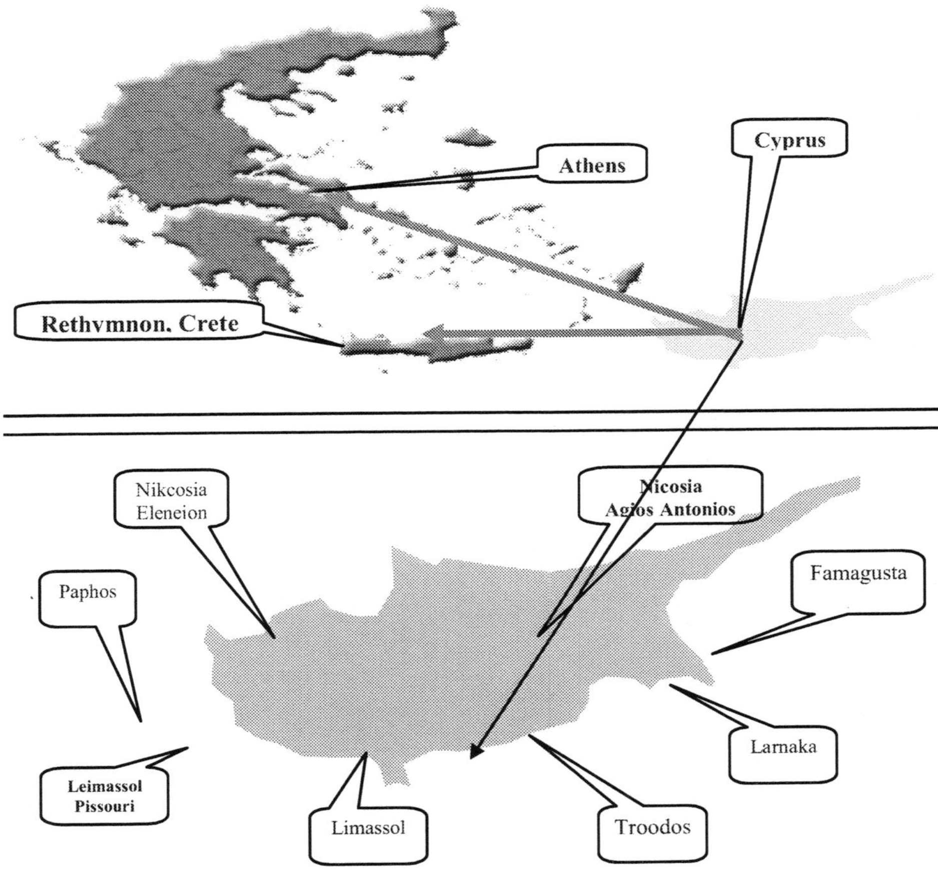

Figure 13. ODYSSEAS 2000-2004 IVC project: 8 elementary Schools from Cyprus connected with 2 elementary Schools from Greece.

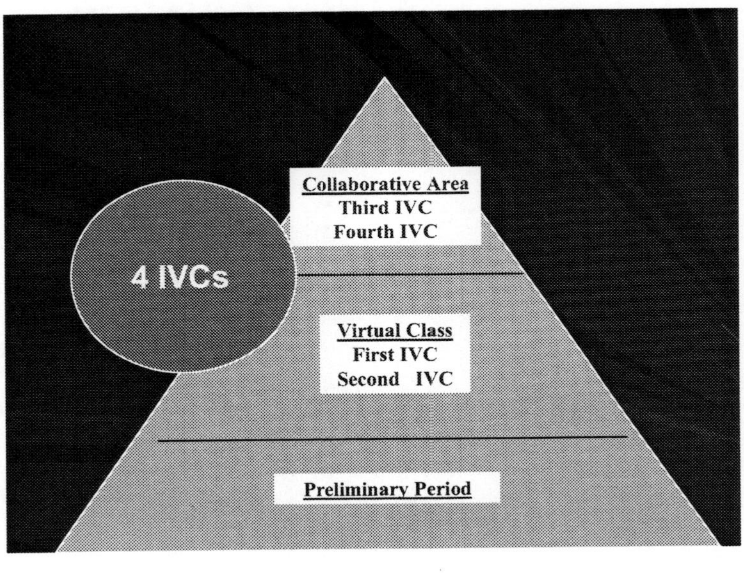

Figure 14.The Stages of the Instructional Methodology.

9.3. The Communication Model Methodology

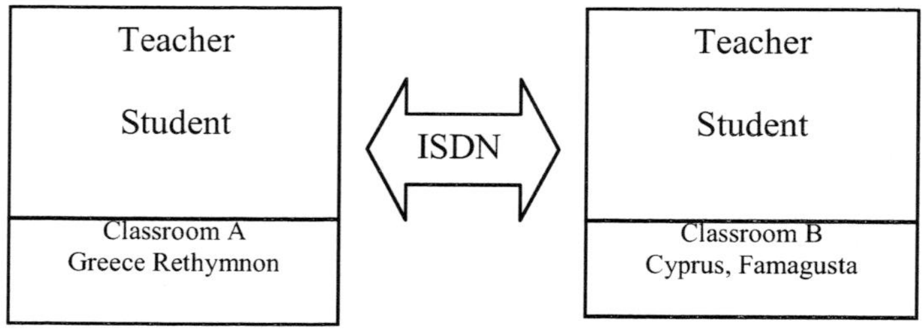

Figure 15. The communication model.

9.4. The Classroom Architecture Model (Anastasiades, 2003b. 2006)

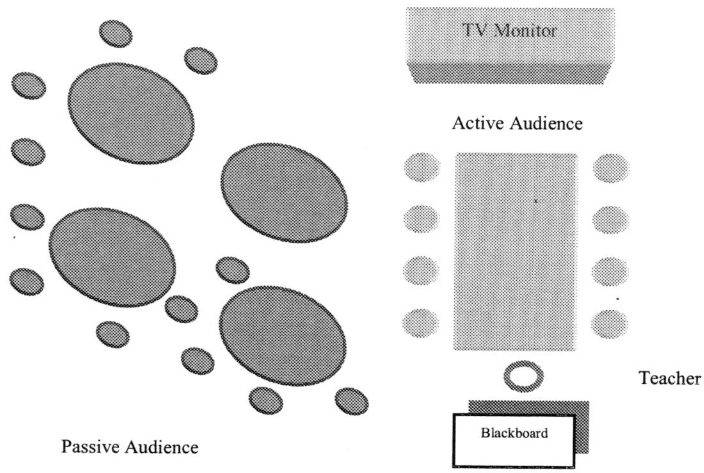

Figure 16. The classroom architecture model.

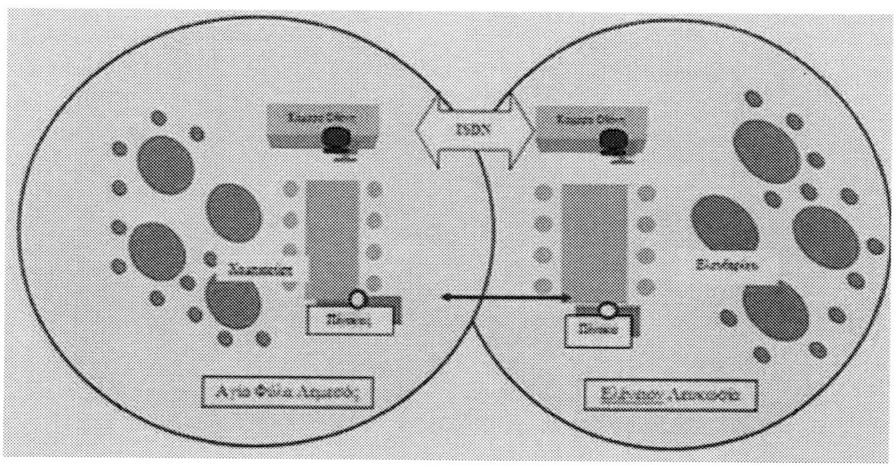

Figure 17. The classroom architecture model: Theory and Practice.

The Virtual Classroom.

Picture 4. Classroom Architecture in Practice.

9.5. ODYSSEAS 2000-2007 IVC Activities

Period February- June 2002: IVCs between Elementary Schools in Cyprus

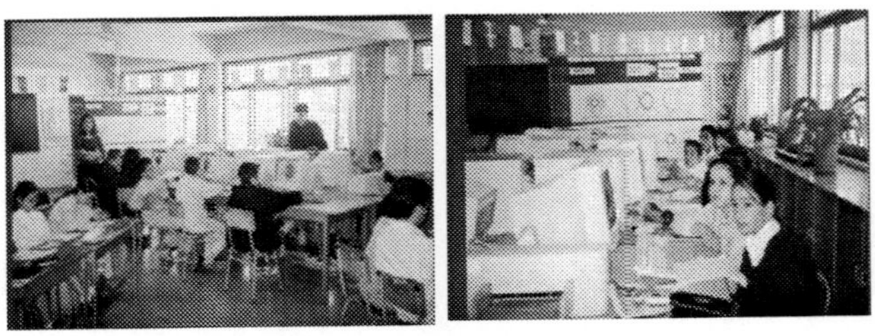

Picture 6. Cyprus 2002 (IVC between Limassol (Agia Fyla) and Larnaka (Zinon). Course: Mathematics).Features: 3 IVCs / 80 minutes each, 4 Teachers, 50 students (Feb- june, 2002).

Picture 7. Cyprus 2004 (IVC between Nicosia (Agios Antonios) and Limassol (Agia Fyla). Cross Curricular Thematic Approach: Olympic Games Athens 2004. Features: 4 IVCs / 80 minutes each, 4 Teachers, 45 students (Feb- june, 2004).

Period February- June 2004: IVCs between Elementary Schools in Crete (Rethymnon) and Cyprus – (Augorou-)

1. Preparing

2. Connecting…

3. Local Dance form Cyprus

4. Local Dance from Crete

5. The first dialogue

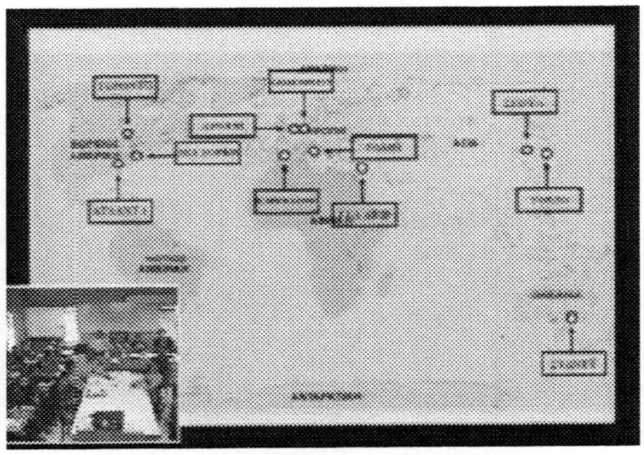

6. PowerPoint presentation

7. Join Collaborative Activity (Pazl)

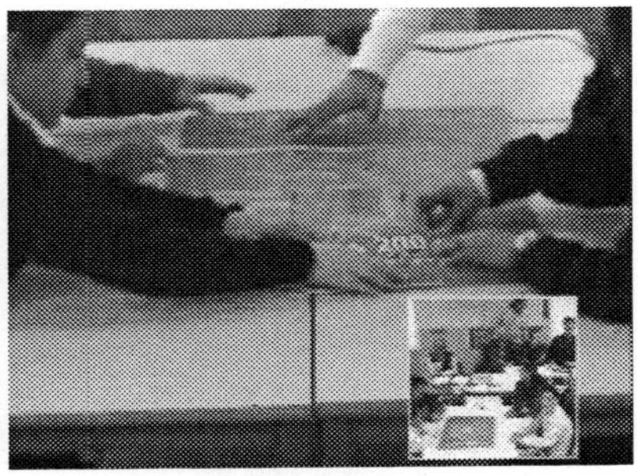

8. and the result

9. Puppet theatre

10. Paintings

11. Calendar

12. Good bye event

Cross Curricular Thematic Approach: Olympic Games Athens 2004 (Special Olmpics) Features: 4 IVCs / 80 minutes each, 4 Teachers, 54 students (Feb- june, 2004).

Period February- June 2006: IVCs between Elementary Schools in Crete (Rethymnon) in Athens and the Island of Aigina

Cross Curricular Thematic Approach: The Sea

Features: 6 elementary clasrooms - 4 IVCs / 80 minutes each

14 Teachers, 180 students (Feb- june, 2006)

Picture 8. Introductory IVC Activities.

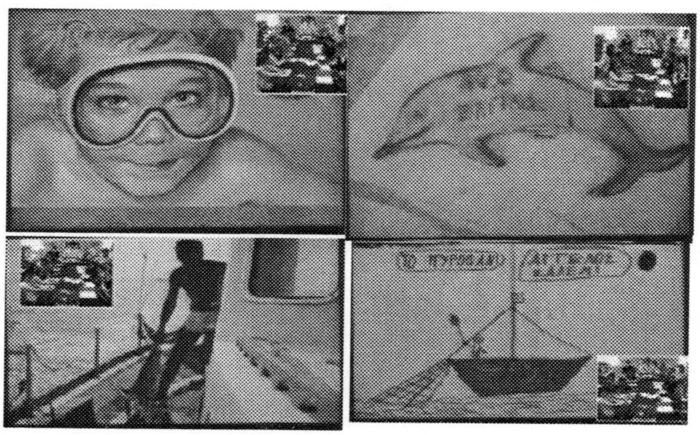

Picture 9. Join Collaborative Activities.

Picture 10. The IVC Room of the Department of Education in the University of Crete.

9.6. The Technological Infrastructure

The main components of the IVC Room of the Department of Education in the University of Crete are as follows:

1. Videoconference System H.320/H.323
2. Interactive Whiteboard Facilities

This device attaches to any whiteboard (up to 4'x8' in size), connects to our PC and when used with a projector, allows us control over our desktop applications and documents directly from the board. Without a projector the interactive whiteboard enables us to digitally capture notes or drawings that can be saved, shared, and/or integrated into other material.

3. Streaming media encoder/server
4. Unidirectional Condenser Boundary Microphone

9.7. The Organization and Support Model

This model fixes the roles that will play the partners of project 'ODYSSEAS'.

The Ministry of Education and Culture of Cyprus has the responsibility of the schools, teachers and students that participate in the project. The department of Computer Science of the University of Cyprus undertakes the scientific co-ordination of the project. In this framework it had the responsibility of instructive methodology development, and the evaluation process, the geographic dimension of the program, the planning of action of dissemination etc. The Bank of Cyprus will have the responsibility for supplying the selective schools of Cyprus with the suitable technological equipment. At the same time it will undertake the economic support of the program. The organizational and support model constituted 3 committees. The monitoring committee, composed of certain representatives of institutions, has the general monitoring responsibility of the project. The research committee has the planning and implementation responsibility of the project. Finally, the organizational committee has a general supporting role. Dr. Panagiotes Anastasiades has the scientific responsibility for this project.

10. Conclusion

The new advanced learning Internet technologies and especially synchronous applications such as videoconferencing will modify the current pedagogical approaches to meet an appointment with the needs of the new virtual learning environment. The proposed methodology as a result of our research since 2000 aims to develop a holistic pedagogic model, which will constitute the necessary theoretical basis and will mark out the framework where the IVC technologies will be contained.

REFERENCE

Abbott, L., et al., (1993). Videoconferencing and Distance Learning. Faculty of Education and Department of Adult and Continuing Education. *University of Ulster*, Ireland.

Ackermann, F. (1996). Participants Perceptions on the role of facilitators using Group Decision Support Systems, *Group Decision and Negotiation*, *5*, 93-112.

ADEC Guiding Principles for Distance Learning and Teaching, (1999). American Distance Education Consortium. (http://www.adec.edu/admin/papers/distanceteaching_principles.html)

Alexander, W., Higgison, C. & Mogey, N. (Eds.) (1999). Videoconferencing in teaching and learning: Case studies. LTDI and TALiSMAN, Institute of Computer-Based Learning, *Heriot-Watt University Edinburgh*, SCT. Retrieved from the World Wide Web June 4, 2004 from www.icbl.hw.ac.uk/ltdi.

Ally, M., (2004). Foundations of educational theory for online learning, in T. Anderson, & E. Fathi, (Eds), *Theory and Practice of Online Learning*, Athabasca University.

Anastasiades, P. (2002a), The Theory of Information Reversal, *Computers and Society, ACM Special Interest Group on Computers and Society (ACM SIGCAS)*, *32(2)*, 10-16.

Anastasiades, P. (2002b). Towards the Global Information Society: The Enactment of a Regulatory Framework as a Factor of Transparency and Social Cohesion, *Lecture Notes in Computer Science Springer-Verlag Berlin Heidelberg*, *2510*, 527-535.

Anastasiades P. (2002c). «Virtual Universities: A Critical Approach». *Proceedings of the International Conference on Computers in Education (ICCE)*, December 3-6 2002, Auckland, New Zealand, College of Business, Massey University, AACE (Association for the Advancement of Computing in Education).

Anastasiades P. (2003a). Implement a Hybrid Learning Environment in Higher Education: A Pilot Methodology for Teaching the Course "ICT and Distance Learning" at the University of Crete. *The International Journal of Learning*, volume, *10*, 297-304, Common Ground, The University Press.

Anastasiades, P. (2003b). «Distance Learning in Elementary Schools in Cyprus: The evaluation Methodology and Results ». *Computers & Education*, *volume. 40*, no. 1, 17-40(24) *Elsevier Science*, January, 2003.

Anastasiades, P., Cwnstantinou, C. P., Sevastidou, A., Eleutheriou, A., Philipidou, Ch., Xristophoridis, M. & Axilews, M. (2003c). «Development Of Children's Investigative Skills Through Collaboration at aDistance Between Two Elementary Schools In Cyprus». *Proceedings of Sixth International Conference On Computer Based Learning in Science*, (CBLIS). July 5-10, 2003, University of Cyprus, Nicosia, Cyprus.

Anastasiades, P., Christou, C., Hambiaouris, C., Georgiou, S., Papadopoulou, E. & Papachristou, K. (2003d). Teaching Mathematics through Collaboration at a Distance between two Elementary Schools in Cyprus». *Proceedings of the Sixth International Conference on Computer Based Learning in Science (CBLIS)*. July 5-10, 2003, University of Cyprus, Nicosia, Cyprus.

Anastasiades, P. (2003e). Building up a pedagogical model for synchronous distance learning courses. In J. A. Chambers, (Ed.), Selected Papers Publication from the 14th International Conference on College Teaching and Learning, April 2003. Jacksonville, FL: Florida Community College at Jacksonville.

Anastasiades, P. (2005). Synchronous Vs Asynchronous Learning? Principles, Methodology and Implementation Policy of a Blended Learning Environment for Lifelong Learning, at the University of Crete. In Proceedings of World Conference on Educational Multimedia, *Hypermedia and Telecommunications*, 2005, (2166-2172). Norfolk, VA: AACE.

Anastasiades, P. (2006). «Interactive Videoconferencing in Lifelong Learning: Methodology and Implementation Policy at the University of Crete (E.DIA.M.ME)» Proceedings of the Diverse 2006, *6th International Conference on Video and Videoconferencing in Education*, Glasgow Caledonian University, Scotland, Glasgow, 5-7 July, 2006.

Anderson, T. (2004). "Toward a theory of online learning" in T. Anderson, & E. Fathi, (Eds), *Theory and Practice of Online Learning*, Athabasca University, ISBN 0-919737-59-5

Andrews, T. & Klease, G. (1998). Challenges of multisite videoconferencing: The development of an alternative teaching/learning model. *Australian Journal of Educational Technology*, *14(2)*, 88.

Angiolillo, J. S., Blanchard, H. E., Israelski, E. W. & Mané, A. (1997). Technology constraints of video-mediated communication. In K. E. Finn, A. J. Sellen, & S. B. Wilbur, (Eds.), *Video-mediated communication*, (51-74). Mahwah, NJ: Lawrence Erlbaum Associates.

Arnold, T., Cayley, S. & Griffith, M. (2002). Videoconferencing in the classroom: *communications technology across the curriculum.*: Devon Curriculum Services.

Bates, A. W. (1995). Technology: Open learning and distance education. New York: Routledge.

Berge, Z. & Collins, M. (1995a). Computer mediated Communication and the Online Classroom in Higher education, *Computer –Mediated Communication Magazine*, vol.2, no.2.

Berge, Z. (1995), The Role of the Online Instructor/Facilitator, *Educational technology*, *35 (1)*, 22-30.

Berge, Z. L. & Mrozowski, S. (2001). Review of Research in Distance Education 1990 to1999.*American Journal of Distance Education*, *15(3)*.

Berge, Z. & Collins, M. (1995). Computer-Mediated Communication and the Online Classroom in Distance Learning, *Computer-Mediated Communication Magazine (CMC)*, *2(4)*.

Bernard, R., Rubalcava, B. & Pierre, D. (2000). Collaborative online distance learning: issues for future practice and research, *Distance Education*, *21(2)*, 260-277.

Bonk, C., Malikowski, S., Angeli C. & Supplee, L. (1998). Holy Cow: Scaffolding Case-Base "Conferencing on the Web", with preservative teachers, San Diego: *American Educational Research Annual Meeting*.

Brown, S. (2001). Views on Videoconferencing Higher Education and Research Opportunities in the UK (HERO), March Issues- http://www.hero.ac.uk/inside_he/archive/views_on_videoconferencin883.cfm?

Bruce, V. (1996). The role of the face in communication: implications for videophone design, *Interactingwith Computers*, *8*, 166-176.

Brusilovsky, P. (1999). Adaptive and Intelligent Technologies for Web-based Education. Kunstliche Intelligenz, *Special Issue on Intelligent Systems and Teleteaching*, *4*, 19-25.

Burge, E. & Haughey, M. (2001). *Visual Learning Technologies, International Perspectives on Practice*. New York: Routledge Falmer.

Carr, W. & Kemmis, St. (2002). Towards a Critical Educational theory – education, *knowledge and action research*, Athens, Kodikas.

Chandler and Hanrahan, (2000) G. Chandler, & P. Hanrahan, Teaching using interactive video: *creating connections*, J. Nurs. Educ. 39 (2000), 73-80.

Chen, Y. & Willits, F. (1998). A path analysis of the concepts in Moore's theory of transactional distance in a videoconferencing environment. *Journal of Distance Education, 13(2)*, 51-65.

Clark, R. (1983). Reconsidering research on learning from media. *Review of educational research, 53(4)*, 445-459.

Clear, A. (1999b). A Collaborative learning trial between New Zealand and Sweden- using Lotus Notes Domino in *Teaching the Concepts of Human Computer Interaction*, In Proseedings of the ITiCSE '99 Conference ACM: Cracow.

Collins, A. (1991). *The role of computer technology in restructuring schools*, Ph Delta Kappan.

Coventry, L. (2000). Videoconferencing in Higher Education, Institute for Computer Based Learning, Heriot Watt University, Edinburg, http://www.man.ac.uk/MVC/SIMA/video3/contents.html

Coventry, L. (1998). Video conferencing in higher education.: *Institute for Computer Based Learning*, Heriot Watt University.

Dallat, J., et al. (1992). Videoconferenccing and the Adult Learner. *Faculty of Education and Department of Adult and Continuing Education*, University of Ulster, 1992, Ireland.

Damankis, M. & Anastasiades, P. (2005). «Life Long & Distance Learning and the Diaspora: Implementing a Virtual Learning Environment at the University of Crete». Themes in Education, Special Issue "*Information & Communication Technologies in Diaspora*. 2005.

De Bra, P., Eklund, J., Kobsa, A., Brusilovsky, P. & Hall, W. (1999). Adaptive Hypermedia: Purpose, Methods and Techniques. *10th ACM Conference on Hypertext and Hypermedia*, 1999, 199-200

Delaney, G., Jacob, S., Iedema, R., Winters, M. & Barton, M. (2004). Comparison of face-to-face and videoconferenced multidisciplinary clinical meetings. *Australasian Radiology, 48(4)*, 487-492.

Dick, W. & Carey, L. (1990). *The systematic design of instruction* (3rd ed.). Glenview, IL: *Scott, Foresman*, and Company.

Digital Bridges: K-12 videoconferencing, Note: another very good resource. Its "Teacher's Guide to Videoconferencing" is valuable and comprehensive, with emphasis on phased approach to planning, delivery and evaluation. A list of reference literature could be updated (the latest referenced articles were from 1999). http://www.netc.org/digitalbridges/vc/index.html.

Ertl, B., Reiserer, M. & Mandl, H. (2003). Cooperative Learning in Videoconferencing: The Influence of Content Schemes and Cooperation Scripts on Shared External Representations and Individual Learning Outcomes (Research report No. 154). Munich, Germany: Ludwig-Maximilians-University, *Department of Psychology, Institute for Educational Psychology*. Research report No. 154, March 2003.

Ertl, Bernhard und Kopp, Birgitta und Mandl, Heinz, (2005). Supporting Collaborative Learning in Videoconferencing using Collaboration Scripts and Content Schemes

(Forschungsbericht Nr. 176). München: Ludwig-Maximilians-Universität, *Lehrstuhl für Empirische Pädagogik und Pädagogische Psychologie*, Internet, ISSN 1614-6336

Fillion, G., Limayem, M. & Bouchard, (1999). Videoconferencing in distance education: a study of student perceptions in the lecture context. *Innovations in Education and Training International, 36*, 302.

Fincher, S. & Petre, M. (1998). *Project-Based Learning Practices in Computer Science Education, FiE Conference*, 1185-1191. IEEE, November 1998.

K. E., Finn, A. J. Sellen, & S. B. Wilbur, (Eds.). (1997). Video-mediated communication. *Mahwah*, NJ: Lawrence Erlbaum Associates.

Fischer, F. & Mandl, H. (2003). Being there or being where? Videoconferencing and cooperative learning, in: H. V. OOSTENDORP (Ed.) *Cognition in a Digital World*, 205–223 (Mahwah, NJ,Erlbaum).

F., Fischer, I., Kollar, H. Mandl, & J. M. Haake, (Eds.) (in press). Scripting computer-supported collaborative learning. *Cognitive*, computational and educational perspectives. New York: Springer.

Fitzgerald, M. (2000). *Educational Media & Technology Yearbook* 2000, Volume 25. Englewood, CO:Libraries Unlimited.

Frey, K. (1998). Η μέθοδος Project (μτφ. Κλ. Μάλλιου). Θεσσαλονίκη: Κυριακίδης.

Gardner, H. (1993). Multiple Intelligence Basic Books, New York.

C. Gibson, & S. Cohen (Eds.), *Virtual Teams That Work: Creating Conditions for Effective Virtual Teams*, San Francisco, CA: Jossey-Bass/Wiley, 2003.

Gowan, J. A. & Downs, J. M. (1994). Video conferencing human-field study. *Information and Management, 27*, 341-356.

Gratton, L. (1998). Video conferencing.: *The Teaching and Learning Centre*, University of New England.

Greenberg Alan, (2006). Taking the Wraps off Videoconferencing in the U.S. *Classroom A State-by-State Analysis Wainhouse Research*, July, 2006.

Gürer, D., Kozma, R. & Millán, E. (1999). Impact of shared applications and implications for the design of collaborative learning environments. In S. Lajoie, (Ed.), *Proceedings for the ninth world conference on artificial intelligence in education* (AI-ED99, 439-445). Amsterdam: IOS Press.

Gustafson, K. L. & Powell, G. C. (1991). Survey of instructional development models with an annotated ERIC bibliography (2nd ed.). Syracuse, NY: ERIC *Clearinghouse on Information Resources.*

Hayden, K. & Hanor, J. (2002). Videoconferencing: A Tool For Collaboration And Professional Development. In Proceedings of World Conference on Educational Multimedia, *Hypermedia and Telecommunications*, 2002, (725-726). Norfolk, VA: AACE.

Hayden, K. L. (1999). Videoconferencing in K-12 Education: A Delphi study of characteristics and cr itical strategies to support constructivist learning experiences. Online [http://hale.pepperdine.edu/~kahayden/dissertation.html]

Hearnshaw Dave, (2000). *Effective desktop videoconferencing with minimal network demands British Journal of Educational Technology, 31(3)*, 221-228.

Hearnshaw, D. (1999). 'Capitalising on the Strengths and availability of Desktop Videoconference', *Active Learning, 9*, 52-59.

Heath, M. & Holznagel, D. (2002). Interactive Videoconferencing: A Literature Review [PDF]. Northeast and the Islands Regional Technology in Education Consortium. Available:http://neirtec.terc.edu/k12vc/resources/necc_draft_062002.pdf

Heath, M. J. & Halznagel, D. (2002, October). Interactive videoconferencing: a literature review. *Prepared for the K-12 National Symposium for Interactive Videoconferencing Dallas*, TX.

Holmberg, B. (1986). *Growth and Structure of Distance Education*, Croom Helm, London.

Hotcomm, (2003). *Synchronous tools and the emerging online learning model.* Retrieved April 18, 2004, from http://www.hotcomm.com/tec/dlwp.pdf

http://www.kn.pacbell.com/wired/vidconf/description.html#what

Hu, C., Wong, A. F. L., Sharpe, L., Crawford, L., Gopinathan, S., Khine, M. S. & Moo, S. N. (2002). Building a Learning Community via VideoconferencingProceedings of the *International Conference on Computers in Education*, (ICCE'02) 0-7695-1509-6/02 $17.00 © 2002 IEEE.

Johansen, D. H. (1999). Designing constructivist learning environments. *eideberg*, FRG:Spriner- Verlag.

IDE,(1996). Three Models of Distance Education. University of Maryland University College. Available online at: http://www.umuc.edu/IDE/modeldata.html

Imel, S. (1994). Guidelines for working with adult learners: *ERIC digest number 77.* Retrieved June 21, 2003 from the ERIC database on the World Wide Web: http://www.ed.gov/databases/ERIC_Digests/ed299456.html

Keegan, D. (1993). *Theoretical principles of distance education.* London: Routledge.

Keegan, D. (1993). *Theoretical principles of distance education.* London: Routledge.

Kerrey, B. & sakson, J. (2001). 'The Power of the Internet for Learning: Moving from Promise to Practice', *Report of the Web-based Training Educational Commision*, Washington DC, 55.

Khan, S. & irata, A. (2001). 'Moving Videoconferencing to IP Networks', *e Learning, 2*, 29-32.

Kingham, S. (2002). *CSIRO case study: facilitating collaboration using video over IP conferencing*. Canberra: Questnet. 2003.

Kinshuk, Hong, H., Patel, A., Oppermann, R. & Russell, D. (2001). Role of Human Teacher in Web-based Intelligent Tutoring Systems. *Journal of DistanceLearning, 6(1)*, 26-35.

Kostoula, N. & Makrakis, (2006). *Intercultural and Education*, Athens.

Knipe, D. & Lee, M. (2002). The quality of teaching and learning via videoconferencing *British Journal of Educational Technology, Volume 33*, Issue 3, 301-311.

Knoll, M .(1997). The Project Method, *Journal of Industrial Teacher Eduction*, vol(34)3.

Knowles, M. S. (1970). The Modern Practice of Adult Education: *Andragogy Versus Pedagogy*, Association Press, New York.

Kozma, R. B. (2001). "Counterpoint theory of learning with media", in R. E. Clark, (Ed.), *Learning from media: Arguments, analysis, and evidence*, 137-178, Greenwich, CT: Information Age Publishing Inc.

Latchem, C. (2002). 'ICT-based Learning Networks and Communities of Practice', *Media and Education, 8, National Institute of Multimedia Education*, (NIME), 1-13.

Lawrence, B. H. (1995). Teaching and learning via videoconference: the benefits of cooperative learning. *Journal of Educational Technology Systems, 24(2)*, 145-149.

Lea, M. & Nicoll, K. (2002). Distributed Learning, *Social and Cultural Approaches to Practice*. New York: Routledge Falmer.

Lian, A. (2000). Knowledge transfer and technology in education: toward a complete learning environment. (http://ifets.ieee.org/periodical/vol_3_2000/v_3_2000.html)

Lionarakis, A. (1998). Polymorphic education: A pedagogical framework for open and distance learning, In Proceedings of EDEN Conference Universities in a Digital Era – transformation, *innovation and Tradition*, Italy: University of Bologna.

Lionarakis, A. (2006). Open and Distance Learning: *Theory and Practice, Probobos*, Athens.

Lockwood, F., (1995). Open and distance learning today. London: Kogan Page.

Martin, M. (2005). Seeing is believing: *the role of videoconferencing in distance learning British Journal of Educational Technology, 36(3)*, 397-405.

Makrakis, V. (2000). Hypermedia and Education: A constructivism approach. *Metaxmio*, Athens.

Marilyn, J. Heath and Donald Holznagel, Kirk deFord, & Victoria Dimock, K. (2002). Interactive videoconferencing: A Literature Review and A Policy Issues Review, Prepared for the k–12 *National Symposium for Interactive Videoconferencing*, Dallas, Texas •

Martin, M. (2000). *Videoconferencing in teaching and learning—case studies and guidelines*. Western Education and Library Board.

Massialas, B. (1989). *Claims and Counterclaims of the educational Uses of Microcomputers*, Third International |Piadgogical Conference ΄Athens, Greece

Mathiews, B. (1992). *The social Issues in information technology for teacher education, 1(2)*, 201-213.

Matsagouras, H. (2002). *A cross curricular thematic approach in Education*. Grigoris, Athens

Maxim, G. (1999). Social Studies and the Elementary Schools Child. *Upper Saddle River*, NJ: Prentice Hall.

Merrill, D., Drake, L., Lacy, M. & Pratt, J. (1996). Reclaiming Instructional Designing, *Educational Technology, 36(5)*, 5-7.

Mitchell, J. et. al (1993). Video-Conferencing in Higher Education in Australia. Canberra: AGPS.

Moore, M. G. (1977). *On a theory of independent study*. Hagen: Fernunivrsiat (ZIFF).

Moskal, P., Martin, B. & Foshee, N. (1997). Educational technology and distance education in Central Florida: An assessment of capabilities. *American Journal of Distance Education, 11(1)*, 6-22.

Motamedi, V. (2001). *A critical look at the use of videoconferencing in United States distance education. Education, 122*, 386-394.

Newlands, A. (1998). *The Effects of Computer Mediated Communication on the Process of Communication and Collaboration*. PhD thesis, University of Glasgow, Human Communication Research Centre, Department of Psychology.

Paulsen, M. F. (1998). Session on Online Teaching Strategie", *Session for the Teaching over the Web Conference*, University System of Georgia, May.

Paulsen, M. F. (2003). Online education and learning management systems. *Global e-learning in a Scandinavian perspective*, Oslo: NKI Forlaget.

Peters, O. (1998). Learning and teaching in distance education: analyses and interpretations from an international perspective. London: Kogan Page.

Peters, O. (2004). *Distance Education in Transition – New Trends and Challenges*, Bibliotheks und Informationssystem der Universitat Oldenburg, *Centre for Distance Education*, 4th edition, 129.

Pitcher, N., Davidson, K. & Goldfinch, J. (2000). Videoconferencing in higher education. *Innovations in Education and Training International*, *37*, 199.

Raptis, & Rapti, (2004). *Teaching and learning towards the information age*, Athens.

Razzouk, R. & Lehman, J. (2002). Video Conferencing as a Tool to Link Colleges of Education with K-12 Schools: A P3T3 Project Initiative. In Proceedings of Society for *Information Technology and Teacher Education International Conference*, 2002, (2424-2425). Norfolk, VA: AACE.

Reed, J. & Woodruff, M. (1995). An introduction to using videoconferencing technology for teaching, *The Distance Educator Newsletter*, Fall, available online: http://www.kn.pacbell.com/wired/vidconf/Using.html Scott, T D and Pitcher, N (1997) SUMSMAN: Scottish

Reed, J. & Woodruff, M. (1995). Videoconferencing. *Using Compressed Video for Distance Learning*, The Distance Educator Newsletter.

Reeves, T. C. (1997). *Evaluating What Really Matters in Computer-Based Education.* University of Georgia. Available: http://www.oltc.edu.au/cp/refs/reeves.htm (21/4/2006).

Roberts, J. (1996). The Story of Distance Education: A Practioner's Perspective, *Journal of the American society for information science, 47(11)*, 811-815.

Robinson, A. (1997). Adult learning by videoconferencing. In Field, J (ed.) Electronic Pathways. *Adult Learning and the New Communication Technologies*, National Organisation for Adult Learning (NIACE), Leicester.

Robyn Smyth, (2005). Broadband videoconferencing as a tool for learner-centred distance learning in higher education. *British Journal of Educational Technology, 36(5)*, 805-820.

Ruippo, M. (2002). *Music studies in synchronous learning environments.*Retrieved April 18, 2004, from http://www.tukkk.fi/tutu/Vision/Papers/MattiRuippo.pdf

Sauve, V. L. (2000). *Issues, Challenges and Alternatives in Teaching Adult ESL*, Don Mills, On: Oxford University Press.

Scardamalia, M. (2002). *Collective cognitive responsibility for the advancement of knowledge.* In B. Smith, (Ed.) *Liberal education in a knowledge society*, (67-98). Chicago: Open Court. Διαθέσιμο στη διεύθυνση: http://ikit.org/fulltext/2002 CollectiveCog.pd (4/9/ 2006).

Schweizer, K., Paechter, M. & Weidenmann, B. (2003). Blended learning as a strategy to improve collaborative task performance, *Journal of Educational Media, 28*, 211-224.

Sharpe, L., Hu, C., Crawford, L., Gopinathan, S., Moo, S. N. & Wong, A. F. L. (2000). 'Multipoint Desktop Videoconferencing As A Collaborative Learning Tool for Teacher Training'. *Educational Technology, 40(5)*, 61-63.

Simonson, M., Smaldino, S., Albright, A. & Zvacek, S. (2003). Teaching and Learning at a Distance, *Foundations of Distance Education*, (2nd edition). Upper Saddle River, NJ: Merril Prentice Hall.

Solomonidou, X. (2006). New Trends on Educational Technology: A constructivism approach. *Metaxmio*, Athens.

Solvie, P. (2003). Videoconferencing: Creating a Space for New Discourse In Literacy and Language Education. In *Proceedings of Society for Information Technology and Teacher Education International Conference 2003*, (2199-2202). Norfolk, VA: AACE.

Squire, K. D. & Johnson, C. B. (2000). Supporting distributive communities of practice with interactive television. *Educational Technology Research and Development*, *48(1)*, 23-43.

Suthers, D. (2001). Collaborative representations: Supporting face to face and online knowledge-building discourse. *Proceedings of the 34th Hawai`i International Conference on the System Sciences*, (HICSS-34), January 3-6, 2001, Maui, Hawai`i (CD-ROM), Institute of Electrical and Electronics Engineers.

Thompson, M. (1994). Speaking personally with Alan Chute, *The American Journal of Distance Education*, *8(1)*, 72-77.

Thompson, R., Rowan, L. & Kindt, I. (1995). Innovations in videoconferencing: embracing technology and joining regions in university teaching, *CQ Journal of Regional Development*, *3(4)*, 57-60.

Twigg, C. (2002). Improving Quality and Reducing Costs: Designs for Effective Learning Using Information Technology. *The Observatory on Borderless Higher Education*, Retrieved: 3/3/2003, from http://www.obhe.ac.uk/products/reports/

Vrasidas, C. & McIsaac, M. (2000). Principles of pedagogy and evaluation of Web-based learning, *Educational Media International*, *37(2)*, 105-111.

Wenger, E. (1998). *Communities of Practice: learning, meaning and identity*. Cambridge: Cambridge University Press.

Woodruff, M. & Mosby, J. (1996). *A brief description of video conferencing. Videoconferencing in the classroom and library.*

In: Videoconferencing: Technology, Impact and Applications ISBN: 978-1-61668-285-9
Editor: Adam C. Rayler, pp. 179-196

Chapter 8

LEARNING FOR K-12 STUDENTS AND TEACHERS: THEORY AND PRACTICE THE CONDITIONS FOR AN EFFECTIVE VC

Panagiotes S. Anastasiades *
University of Crete, Rethymnon, Greece

INTRODUCTION

The organization of a successful VC requires detailed design and preparation, as there are a number of factors to take into consideration. Depending on the VC type, the participants' needs and the set goals we should investigate aspects like the necessary technological equipment, the lay-out of venues (Anastasiades, 2007; Mouzakis et al., 2004), the organizational arrangements and, last but not least, the conditions for a warm and collaborative environment for the participants (Anastasiades, 2008).

OBJECTIVE

In this chapter we present a detailed description of the technological, organizational, designing and social conditions in order to provide a thorough framework for the implementation of successful VCs.

STRUCTURE

The structure of this chapter is presented in figure 1

* Associate Professor, University of Crete, Department of Education, Greece. Email: panas@edc.uoc.gr
Home Page: http://www.edc.uoc.gr/~panas/

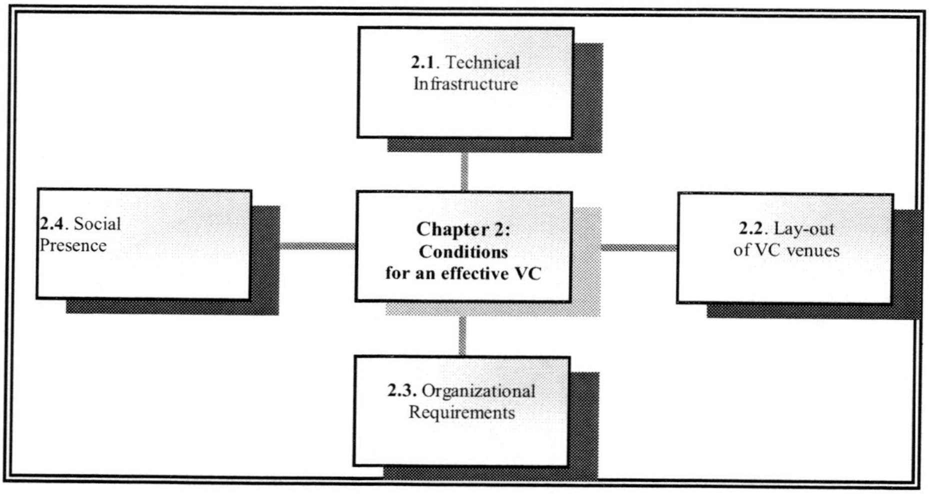

Figure 1. The structure of chapter 8

1. Technical Infrastructure

In order to organize a successful VC, we should ensure that there are the necessary technical requirements – audio, image, data sharing, additional equipment – as well as the necessary communication infrastructure – connection speeds, available bandwidth, etc. (figure 2).

Sound and image play a key role to the successful implementation of a VC and the creation of a warm and friendly atmosphere for the participants. The quality of image and sound depends on the capacity of the equipment (microphones, cameras, speakers, etc.), the available bandwidth, the connection speed of the distant locations and the capacity for coding – decoding audio and image data. Furthermore, the possibility to share data and applications between the distant locations is a significant factor contributing to interactive VCs.

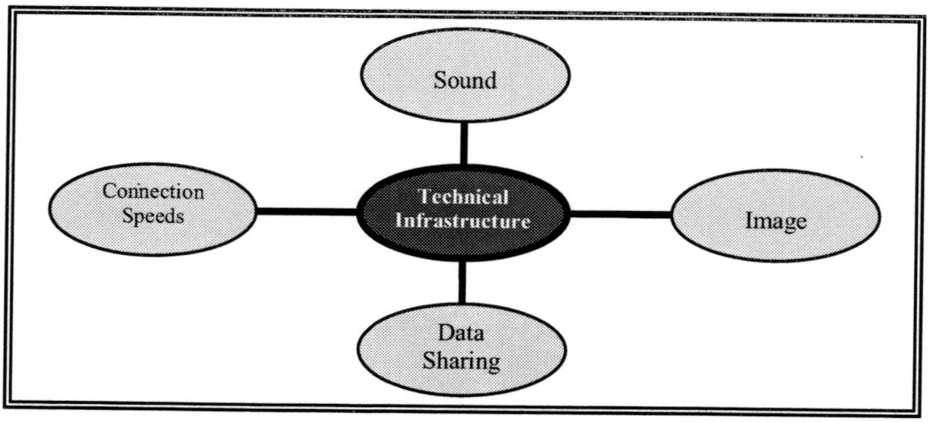

Figure 2. The technical infrastructure for a successful VC

1.1. Sound

From our experience in VCs so far we have concluded that in a VC sound plays a more significant role than image (Videoconferencing Cookbook, 2006). The participants can tolerate brief interruptions in video or frame freeze, but audio problems (frequent pauses, distortion, external noise) are common reasons for the termination of a VC (Balaouras, 2005). The quality of sound in a VC depends on the devices of incoming and outcoming sound, the audio coding and the connection speed.

a. Devices of incoming and broadcast sound

The broadcast sound (what we broadcast to distant locations) is transmitted in real time from the microphones in our venue to the points we are connected.

For One-to-One VCs, it is required that we use a headset. For group VCs, we usually use a high-quality omni-directional microphone (Picture 1.a). In cases of bigger audiences, omni-directional microphones are installed so that they cover the whole venue space (Picture 1.b). The incoming sound (what we receive from the distant locations) is transmitted through the speakers (Picture 1.c), which should be of the capacity to cover the whole venue space.

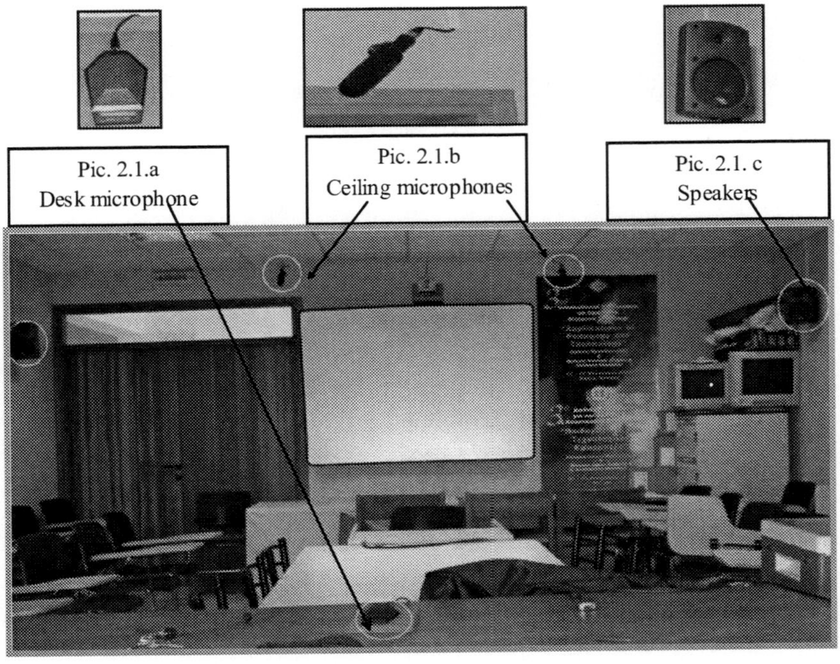

Picture 1. Sound devices in a VC room

Tips and advice

In every case, microphones must not be placed within the speakers' range, because this will result in resonance to the distant location, which can be very annoying during a VC.

b. Audio coding standards

VCs based on the H.320 (ISDN) standard are supported by three audio compression standards: G.711, G.722 and G.728, whereas the VCs on H.323 (IP) standard are supported by G.711, G.722, G.728, G.723 and G.729.

The G.711 standard is the basic audio compression in most VC systems, so that there is equipment compatibility. It provides an average quality of sound (upper frequency limit 3.4 kHz/s, as in a telephone line and occupies 56 – 64 kbit/s of data).

The G.722 standard is an advanced audio coding and provides better quality of sound (upper frequency limit 7 kHz/s and occupies 48/56/64 kbit/s of data). It is the ideal audio compression standard for VCs. The G.722.1 Annex C standard, provides excellent quality of sound (upper frequency limit 14 kHz/s).

The G.728 standard occupies the minimum bandwidth, compressing the signal as much as possible. It supports coding of low-rate transmission (upper frequency limit 3.4 kHz/s and occupies 16kbit/s of bandwidth). However, it does not provide adequate sound fidelity.

c. Connection speed and bandwidth of networks

A common speed at 324Kbps can support high-fidelity sound in a VC.

1.2. Image

If sound fidelity can guarantee the implementation of a VC, the quality of image is crucial for the feeling of real communication and the quality of interaction between participants. The quality of image depends on the features of the VC camera, the Video Coder and Decoder, the quality and bandwidth of the connection and the projection equipment. In particular:

a. Technical features of the VC camera

A VC camera captures the motion in a location to broadcast it to another in real time. The most important component of the camera is the image sensor, which captures 25-30 frames per second. The main features of an image sensor are:

- resolution, namely the pixel count a picture holds. The usual standard is 640x480 pixels for webcams, 720x480 for NTSC cameras and 720x576 for PAL cameras.
- technology: the CMOS sensor is mostly used in webcams (Picture 2), which are economical but provide low-quality images. PTZ cameras (Picture 3) can filter and magnify the image angle and focus either automatically or manually on the speakers depending on the needs.

Picture 2. web camera Picture 3: PTZ camera.

Tips and Advice

In every case, microphones must not be placed within the speakers' range, because this will result in resonance to the distant location, which can be very annoying during a VC.

b. The features of Video Coder – Decoder

A coder – decoder is the main component of the VC system. It defines the band of algorithms applied for the coding and decoding of visual data, so that it can be broadcast via Internet connection to a distant location and achieve communication in real time. The most popular standards nowadays are H.261 and H.263 for image coding, which can support VCs based on H.320 and H.323.

The H.261 Video Codec (1992) is up to now the basic standard which defines the type of image coding, so that the image can be broadcast through networks of average speed. The H.263 Video codec is a variation of H.261 and has been specially designed to transmit visual data through networks of low speed (H.324-GSTN and H.323-IP at 64 – 128 kbit/s). The H.264 is the most recent (2005) Video codec supporting video resolutions (19210x1080). The video resolutions supported by the popular VC standards (H.323) are CIF (352x288 pixels), OCIF (176x144 pixels) and 4CIF (704x576).

c. The quality and bandwidth of network connections

The common speed at 384Kbps can support adequately VCs of limited motion of the participants (e.g. a lecture or a board meeting). For cases where motion is intense, such as in a drama play, higher speeds are required (768Kbps), as well as support by the H.320 (ISDN VC) standard.

d. Projection media

When the image is screened on the PC monitor, its quality is affected by the size and resolution of the monitor (picture 4). On the other hand, when the image is screened through a video projector (LCD or DLP), the quality depends on the technical features of the available equipment.

1.3. Sharing data – applications

In many VCs it is necessary to share data and applications between the distant locations, so that the participants can share files or applications, such as PowerPoint presentations (Picture 6), Word documents, web pages, a Smart board (Picture 7), a document camera (Picture 8), etc.

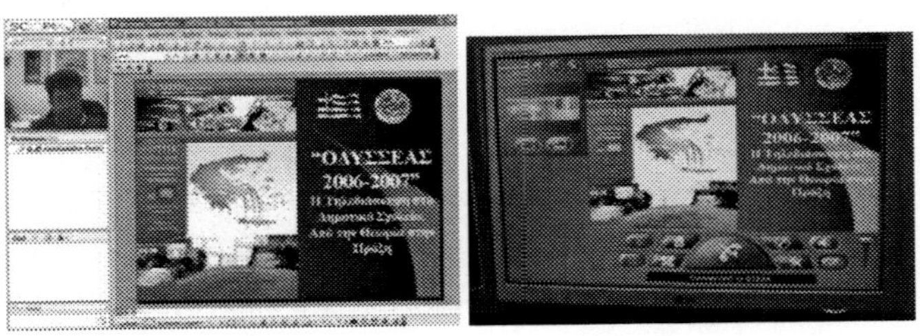

Picture 4. Image on the PC monitor.

Picture 5. Projection via video projector on a screen.

Picture 6. Sharing a PowerPoint presentation in a VC.

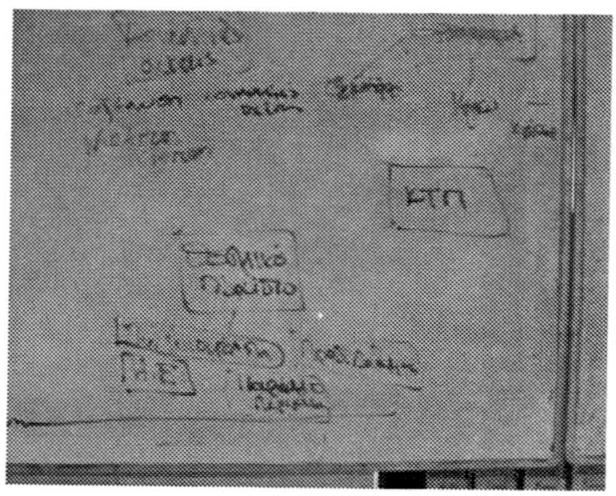

Picture 7. Sharing Image from a White board in a VC.

An electronic board is required in VCs where the speaker wishes to use the board to solve exercises, draw tables, diagrams, etc. It is comprised of one piece, which can be adjusted on any white board (picture 7.a), and the accompanying software.

The Document camera is necessary in VCs where the speaker has to use visual material in a non-digital form, such as a book (picture 8.a), a picture, an appliance, an object, etc.

In the past the exchange of data between the distant locations was achieved via T.120 standard, which supported the sharing of applications and text in real time by defining the protocols for audiographic and audiovisual applications. The basic standard T.120 guaranteed compatibility in systems and did not require knowledge of the technical features of the systems in advance. The T.120 standard inserted data by-passing the transmission channel (H.320, H.323). Thus, only one channel was occupied for transmission of sound, image and data. In the meantime, the low resolution of the broadcast of data had negative effects on the participants' comprehensive requirements (picture 9).

T.120 standard is no longer supported by Windows Vista. One of the newest standards to come out the International Telecommunication Union – Telecommunication (ITU-T) is the H.239 standard. This standard allows video codecs such as H.323 and H.320 to send two video channels within a single video call, which gives the user the ability to show not only people but also data, such as an Excel spreadsheet or a Word document. The screen can show a picture within a picture on a single monitor or separate information on two different screens, one screen for video and another for data (IHETS, 2008).

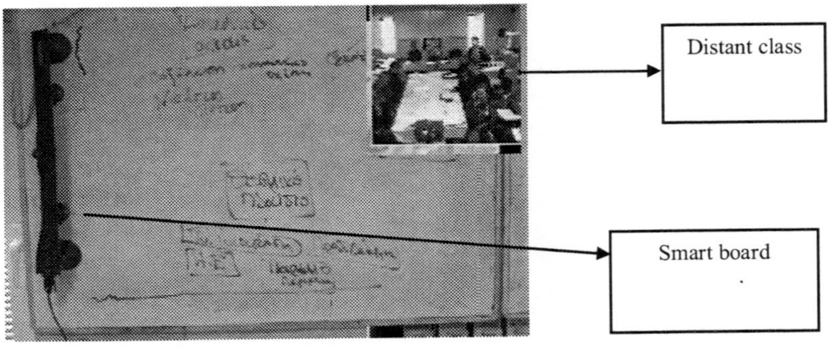

Picture 7.a. The Smart Board.

Picture 8. Sharing image via a Document camera in a VC.

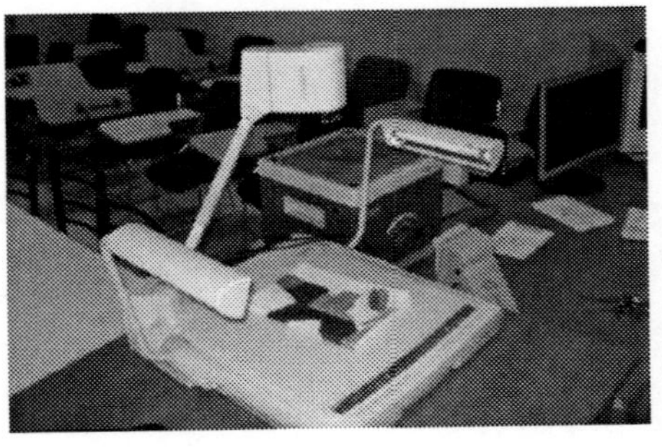

Picture 8.a. The document camera.

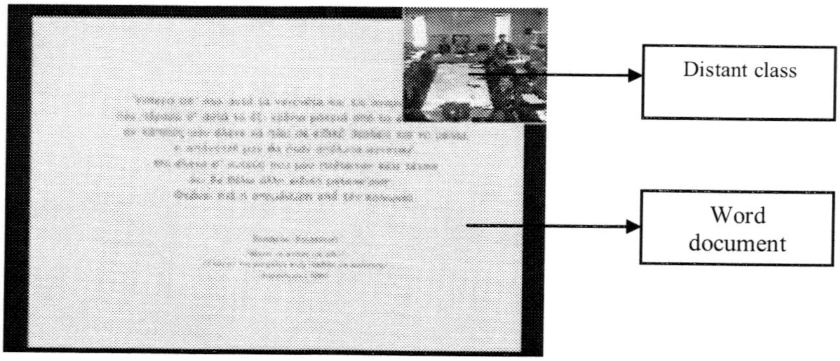

Picture 9. Low resolution in the Word document hampers comprehension.

In order to maximize the benefits of sharing the data services, we should acknowledge that detailed preparation of the specific material is necessary.

> **Tips and advice**
>
> Visual data projected to distant locations through slides using PowerPoint or Document camera must be legible (at least at 24 Font size Arial & Tahoma), by avoiding dense text and complicated graphics (picture 10)

1.4. Connection speeds

The required connection speeds of the distant locations are presented separately for standards H.320 and H.323:

a. Connection speeds for standard H.320 (ISDN VC)

VCs employing the H.320 standard are considered basically effective when connection speed is at 128 Kbps (1 ISDN BRI line).

> **Tips and advice**
>
> In VCs which require high-quality image (video) or include intense motion (Drama play, group collaboration, etc.), a connection speed at 384 Kbps is necessary (3 ISDN BRI lines).

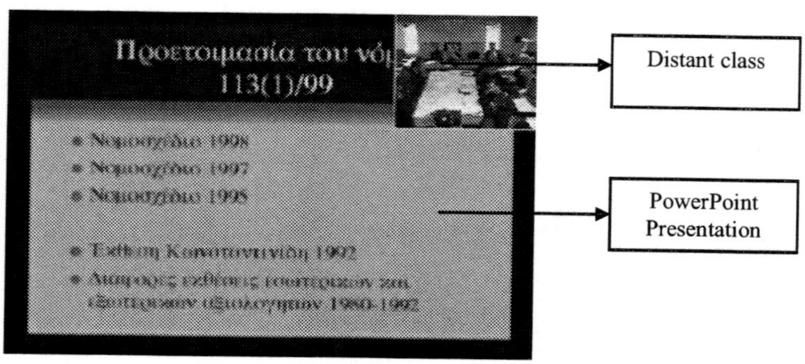

Picture 10. Sharing PowerPoint in a VC.

b. Connection speeds for standard H.323 (IP VC)

VCs applying the H.323 standard are considered effective at connection speed 384 Kbps, while speeds over 768 Kbps are required for high-qubality image and sound.

We should consider that H.323 standard does not guarantee a steady connection speed between distant locations and it burdens network capacity by 30%.

Tip and advice

In VCs based on H.323 standard, when trying to connect at 1024 Kbps, it is essential that we have a capacity of 1350 Kbps, otherwise the quality of image and sound will be low.

Tips and advice

Particularly for ADSL connections (Asymmetric Digital Subscriber Line), we should ensure that the uplink capacity is at least 512 Kbps so that communication is effective.

The following table is a comparative presentation of the features of communication standards in terms of cost, capacity requirements and effectiveness in intense motion.

2. Lay-out of Venue

The lay-out of a VC room is an important factor for its successful implementation. It depends on the type and the purpose of the VC, the participants' needs and the available budget. One-to-One VCs do not require a specific arrangement in the venue, whereas VCs with groups or bigger audience need a detailed lay-out design of the venue.

Table 2.1. Comparative presentation of communication standard features

Communication standards	Cost (connection charge fee)	Capacity requirements	Effectiveness in intense motion
H.320 ISDN Videoconferencing	High	128-512 Kbps	High
H.323 IP Videoconferencing	None	<512 kbps	Low

When arranging the lay-out, we should consider a number of factors, such as the arrangement and the interior design (acoustics, sound and lighting).

2.1. Arrangement of the room

The lay-out of the room largely depends on the number of participants and the purpose of the VC. For small audiences (10 - 15 persons), we usually apply **U** arrangement (figure 2) or **Π** arrangement (figure 3), which are suitable mainly for meetings, boards and round-table conferences).

For medium audiences (20 – 50 persons) the lay-out of the room depends largely on the purpose of the VC. In case of speeches, lectures, presentations, etc., the classroom lay-out is usually recommended (figure 4)

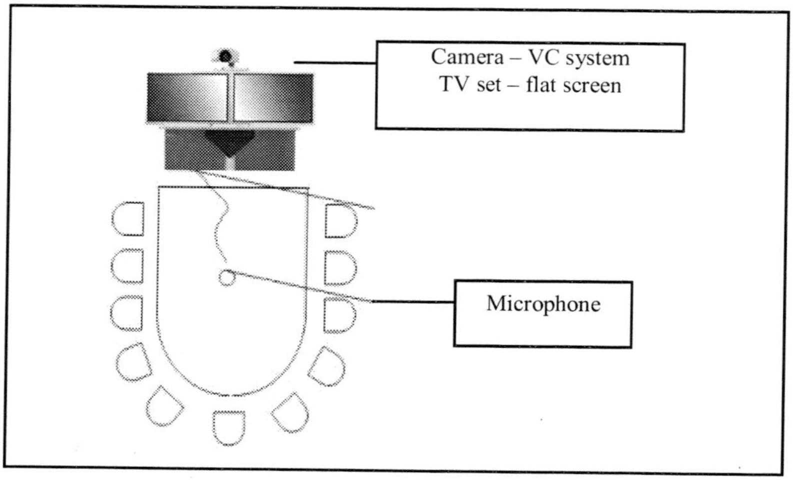

Figure 2. **U** arrangement in a VC room.

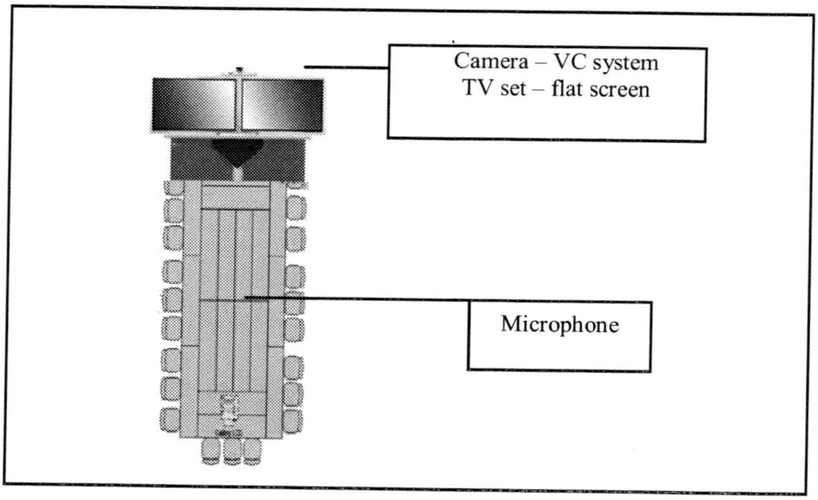

Figure 3. **Π** arrangement in a VC room.

The Classroom lay-out is recommended when medium or bigger audiences attend a lecture or a presentation. Delivery of a speech in a room with this lay-out requires additional arrangements (extra camera, extra projection screen, etc.). For bigger audiences (over 50 persons) the amphitheatre lay-out is recommended (figure 5).

In VC rooms, additional equipment is required for broadcast of sound and image. Installation of extra microphones, speakers, projection screens, flat screens, etc. is necessary.

VC rooms should be equipped to support a wide range of activities, depending on the participants' changing needs. Most times the cost of equipment and the limited availability of venues do not permit the set-up of multiple venues.

U and Π lay-out support VCs of small audiences. Classroom and amphitheater lay-out are convenient for big audiences, but require additional arrangements to support interactive presentations, group collaboration, etc. In the years of our experience in the design and organization of VCs in the University of Crete, and especially at the Department of Elementary Education, the "flexible" lay-out was successfully implemented (Anastasiades. 2003; 2008). The "flexible" lay-out combines Π and amphitheater arrangements (figure 6).

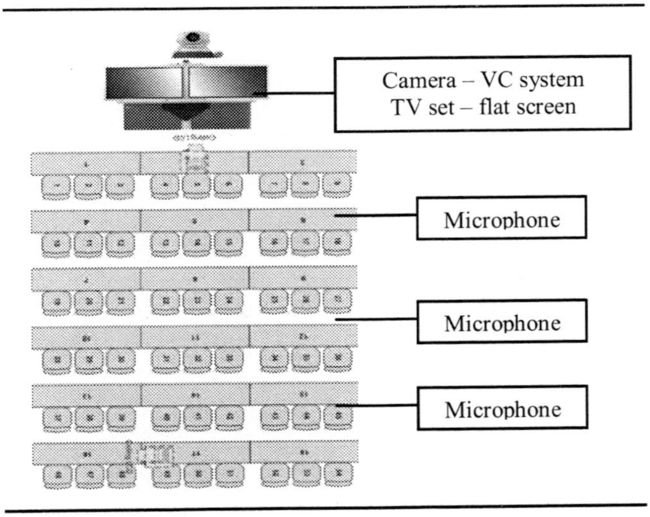

Figure 4. Classroom lay-out in a VC room.

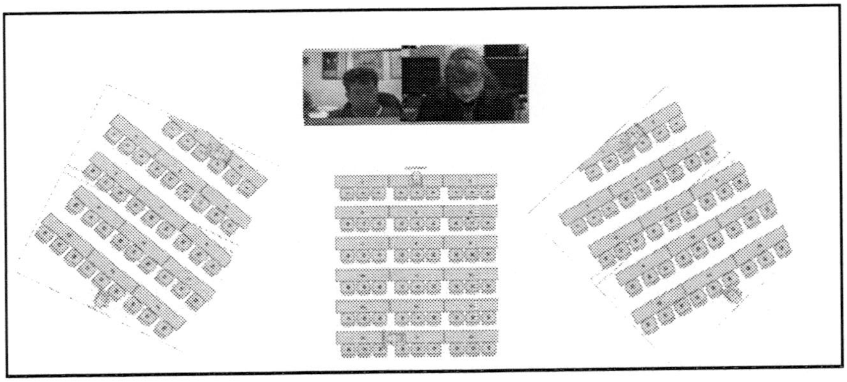

Figure 5. Amphitheater lay-out in a VC room.

-Functionality for participants

The VC camera is installed directly above the projection screen, so that participants can look simultaneously at the distant location (on the screen) and the camera. In the meantime, their visual range may cover the local room. This is especially significant, as it reinforces eye contact, as if in face-to-face communication.

-Functionality for the speaker

The lay-out allows the speaker to address both the local and the distant audience, as if he/she was addressing audience physically present. This considerably helps speakers, as they keep eye contact with both audiences and do not need to take additional action, such as focusing on the extra screen or camera, which can be an embarrassing or irksome task.

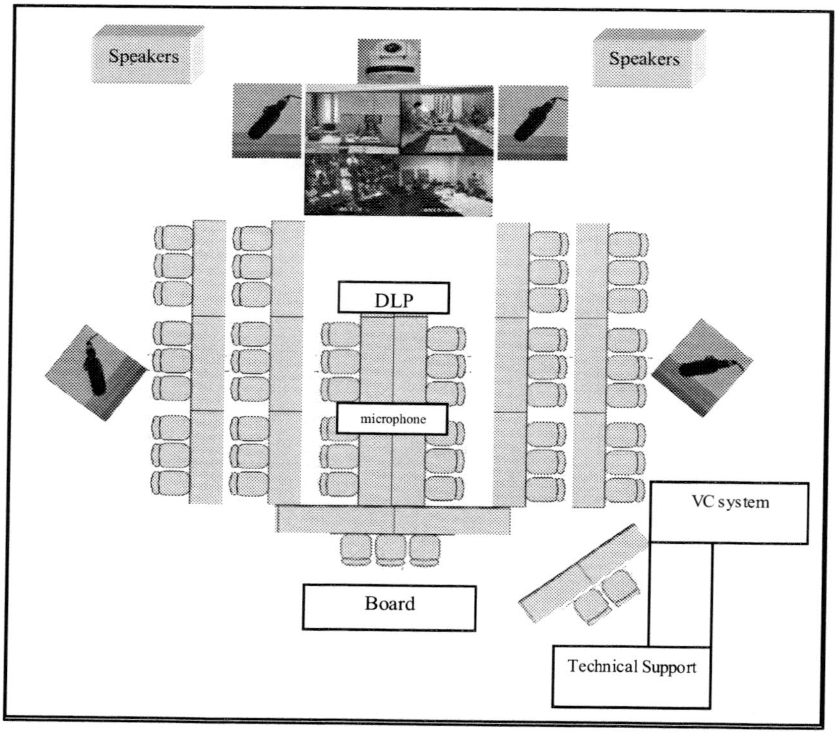

Figure 6. “flexible” lay-out in a VC room (Anastasiades, 2003; 2008).

The advantages of a “flexible” lay-out are the following:

- Versatile Uses

The “flexible” lay-out gives the ability to use the room in different ways:

- meeting of small number of participants (Π arrangement)
- attend a presentation from a distant location (Classroom arrangement)
- presentation to multiple distant locations
- VC for group collaboration, etc.

-Saving resources

Unlike the other lay-outs, the “flexible” arrangement saves technical resources, as there is no need for extra camera, screen or any other additional equipment (screens, video projectors etc). The arrangement of the microphones is adeequate for all the participants.

-Technical support

This lay-out permits the users (for example lecturers who are operating the VC system, or the support team) to make their own technical arrangements on the grounds of their needs.

Tips and advice

All furniture in a VC room (desks, chairs) must be portable, so that every time the arrangement meets the participants' needs. In other words, it should not be fixed on the floor or wired.

2.2. Interior design

The interior design of a VC room includes:

a. Safety

We must conform to general safety rules, and also see to the participants' safety from the system and equipment wiring.

Tips and Advice

Electricity wiring must be mounted and safely away from the participants' space.

b. Colors

Earthly colors are generally recommended for the furniture and the walls, so that the camera records a natural depiction of the participants. It is advisable that we avoid intense colors (red, white), contrasts (white – black) and glossy reflecting surfaces.

Tips and advice

Select soft earthly colors and avoid intense contrasting colors.

c. Lighting

Although most advanced cameras are automated to produce optimum effects, the lighting of the room plays a significant role to the quality of the broadcast image.

Tips and advice

The room must be equipped with light bulbs over 125W

d. Acoustics

The VC room must be away from any noise source. Otherwise, it is necessary to use sound proof walls in order to reduce external noise, which can be reproduced by the microphones and further broadcast to the distant locations. If the ceiling is high, a false roof can considerably enhance acoustics.

Tips and advice

Carpeting the floor, drawing curtains made of soundproof material and limiting glass or wood surfaces can enhance acoustics in a VC room.

A key factor for the broadcast of sound is the type and lay-out of microphones. We should consider the size and lay-out of the room, as well as the number of participants and their activities.

Tips and advice
A combination of ceiling and wireless microphones can be effective for medium-sized VC rooms. For smaller venues, an omni-directional microphone is adequate.

3. Organizational Arrangements

It is necessary to organize the VC in great detail according to the type and needs.

A. Features and Needs of speakers

Each speaker can implement the methods and use the media or tools he/she considers best for a successful presentation. One may wish to use slides to make a PowerPoint presentation, or to demonstrate a book or other print material. Another one may need to navigate the Web, or use audio/image files, etc.

Tips and advice
You should contact the speakers in advance before the VC sessions, so that you provide them with information about the means and techniques they will be using. Explain possible restrictions and necessary customization of the specific conditions. Also, suggest any possible alternative options.

B. Number of participants and Length of the VC

The number of participants must correspond to the capacity of the room. Also, it is advisable that the time length of the VC does not exceed the 60 – 90 mins.

C. Time Zones

We should consider the different time zones of the distant locations that are connected to the VC.

Tips and advice
To schedule VCs between locations of different Time zone, you can consult websites of Time zone Converters (*http://aa.usno.navy.mil* , *www.timezoneconverter.com*).

D. System Check – Testing

A pilot/trial connection should precede a VC to check sound, image and data sharing. For VCs involving multiple distant locations, it is advisable to perform a check simultaneously, so that any possible weaknesses pertaining to the connection of distant points can be spotted and surmounted.

Tips and advice
Make sure that you have the telephone numbers of the technicians at every distant location, so that you can contact them in case of disconnection or inability to connect.

E. Implementation of the VC

During the VC it is essential that there is access to telephone in the room. In addition, there must be easy access to personal information (names and telephone numbers) of the technicians at the distant locations in case of unexpected audio/visual problems.

4. Social Presence

Is face-to-face communication different compared to technology-mediated communication?	

Sproull and Keisler (1986) maintain that the main difference between face-to-face communication and technology-mediated communication is the absence of social context and emotions, resulting from the restrictions of non-verbal conduct.

The VC, as a form of video-mediated communication, allows the participants to communicate verbally and non-verbally from distance (Bondareva, 2005), which, under certain circumstances, can be conducive to the feeling of "Social Presence" among the participants.

The concept of "Social Presence" is nothing new. According to Mehrabian (1969), Social Presence is defined as the outcome of conducts which reinforce both verbal and non-verbal communication (facial expressions, body language, eye contact) and it greatly contributes to the creation of a warm and friendly atmosphere.

Achieving the atmosphere of "social presence" has been the focal point of research from the international scientific community and the researching departments of VC manufacturers. Despite the fact that modern VC systems can support two-way transmission of sound in real time, it appears that they fall short of creating the feeling of interpersonal contact among the participants, equivalent to that in face-to-face communication (Bonk et al., 1998; Collins, 1991; Schweizer et al., 2003).

The inability to socialize, the absence of informal interaction (Chidabaram & Jones, 1993; Rice & Williams, 1984) in combination with the slow flow of information and the complete absence of spontaneity (Campbell, 2000) may transform mediated communication into a self-centred and unfriendly procedure. The participants do not share a common three-dimensional space but can only see what the camera screens, while non-verbal stimuli are ambiguous (Bruce, 1996). In addition, they cannot take advantage of eye consciousness, as, despite visual contact, there is no eye contact with the other participants (Angiolillo et al., 1997).

How can we enhance and reinforce the feeling of "social presence" in a VC?	

Reinforcing the feeling of "social presence" in a VC is connected with the image size of the distant participant (Prussog et al, 1994), the quality – resolution of the broadcast image and the fidelity of undelayed sound (Issacs et al, 1997).

Recent research on "social presence" focuses on the concept of "physical presence", namely the extent to which a person perceives the others as "physically present" during communication with each other (Campbell, 2000). A significant factor to reinforce "physical presence" is to establish the feeling that the participants share a common space (Regenbrecht et al., 2004; Vertegaal, 1999; Nakanish et al., 1998). This fact is important as it challenges the technocrat approach that a VC brings people at different locations into contact. Hence, we are provided with the principles for a new socio-centred approach that the VC environment must create for participants at different locations the feeling of sharing a common space.

The size and quality of the broadcast image in real time, the fidelity of undelayed sound and the reinforcement of the feeling of "sharing a common space" are key factors to consolidate the feeling of "social presence" (figure 7) and point to the future steps in Videoconferencing.

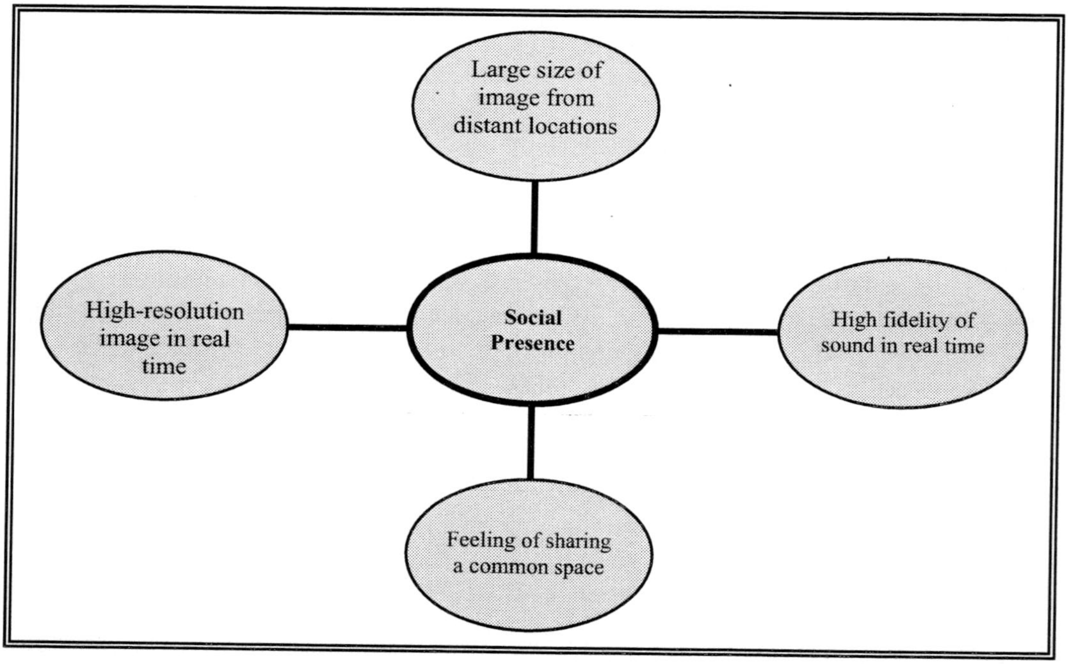

Figure 7. Factors which consolidate the feeling of "social presence" in a VC.

REFERENCES

Anastasiades, P. (2003). Distance Learning in Elementary Schools in Cyprus: The evaluation Methodology and Results. *Computers & Education: volume, 40*, no. 1, 17-40 (24), *Elsevier Science.*

Anastasiades, P. (2007). Interactive VideoConferencing (IVC) as a Crucial Factor in Distance Education: Towards a Constructivism IVC Pedagogy Model under a cross curricular thematic approach. In E. Bailey, (Ed), *Focus on Distance Education Developments*. NY: Nova Science Publishers, Inc.

Anastasiades, P. (2008). Blending Interactive Videoconferencing and Asynchronous Learning in Adult Education: Towards a Constructivism Pedagogical Approach. A Case Study at the University of Crete (EDIAMME). In S. Negash, (Ed), *Distance Learning for Real-Time and Asynchronous Information Technology Education*. NY: Idea Group Inc.

Angiolillo, J. S., Blanchard, H. E., Israelski, E. W. & Mané, A. (1997). Technology constraints of video-mediated communication. In K. E., Finn, A. J. Sellen, & S. B. Wilbur, (Eds.), *Video-mediated communication*, (51-74). Mahwah, NJ: Lawrence Erlbaum Associates.

Bondareva, Y. (2003). Determinants of Social Presence in Videoconferencing. Eindhoven: *Stan Ackermans Institute No. 033.*

Bondareva, Y., Meesters, L. & Bouwhuis, D. (2005). Eye Contact as a Determinant of Social Presence in Video Communication. Retrieved from http://www.hft.org

Bonk, C., Malikowski, S., Angeli C. & Supplee, L. (1998). Holy Cow: Scaffolding Case-Base "Conferencing on the Web" with preservative teachers. San Diego: *American Educational Research Annual Meeting.*

Bruce, V. (1996). The role of the face in communication: implications for videophone design, *Interacting with Computers*, *8*, 166-176.

Campbell, J. (2000). User Acceptance of Videoconferencing: Perceptions of Task Characteristics and Media Traits. In *Proceedings of the 33rd Annual Hawaii International Conference.* Hawaii.

Chidambaram, L. & Jones, B. (1993). Impact of Communication Medium and Computer Support on Group Perceptions and Performance: A Comparison of Face-to-Face and Dispersed Meetings. *MIS Quarterly*, (17, 4), 465-491.

Collins, A. (1991). The role of computer technology in restructuring schools. *Phi Delta Kappan, 73(1)*, 28-36.

Issacs, E. & Tang, J. (1997). Studying video-based collaboration in context: from small workgroups to large organizations. *Video-mediated communication*, 173-197.

Mehrabian, A. (1969). Some referents and measures of nonverbal behaviour. *Behavior Research Methods and Instrumentation, 1(6)*, 205-207.

Nakanishi, H., Yoshida, C., Nishimura, T. & Ishida, T. (1998). FreeWalk: A Three-Dimensional Meeting-Place for Communities. In T. Ishida, (Ed), *Community Computing: Collaboration over Global Information Networks*, 55-89. John Wiley and Sons.

Prussog, A., Miihlbach, L. & Bocker, M. (1994). Telepresence in videocommunications. In *Proceedings of the Human Factors and Ergonomics Society 38th annual meeting*, (25-38).

Regenbrecht, H., Lum, T., Kohler, P., Ott, C., Wagner, M. T., Wilke, W. & Mueller, E. (2004). Using Augmented Virtuality for Remote Collaboration. *Presence: Teleoperators and Virtual Environments*, vol. 13, 338-354.

Rice, R. E. & Williams, F. (1984). Theories Old and New: The Study of New Media. In: R.E. Rice (ed.), *The New Media: Communication, Research and Technology*, 55-80. Beverly Hills, CA: Sage Publications.

Schweizer, K., Paechter, M. & Weidenmann, B. (2003). Blended learning as a strategy to improve collaborative task performance. *Journal of Educational Media, 28,* 211-224.

Sproull, L. & Keisler, S. (1986). Reducing social context cues: Electronic mail in organizational communication. *Management Science, 32,* 1492-1513.

Vertegaal, R. (1999). The GAZE groupware system: mediating joint attention in multiparty communication and collaboration. In *Proceedings of the SIGCHI conference on Human factors in computing systems: the CHI is the limit Pittsburgh.* Pennsylvania: United States ACM Press.

Mouzakis, H., Balaouras, P., Roussakis, I. & Mattheou, D. (2004). Implementation of Synchronous DL Environments in tertiary education. In: M., Grigoriadou, A., Raptis, S. Vosniadou, & H. Kinigos, (Ed), *Proceedings of 4th IT Conference - IT in Education,* (699 -708). Athens: New Technologies Editions.

Videoconferencing Cookbook, (2006). Retrieved June 20, 2006 from: www.vide.net cookbook/cookbook.en/

INDEX

A

D

E

F

G

H

I

J

L

M

P

Q

R

S

T

U

V

W

X

HEALTH CARE REFORM
(Part 2)

JOINT HEARINGS

BEFORE THE

SUBCOMMITTEE ON HEALTH AND THE ENVIRONMENT

AND THE

SUBCOMMITTEE ON COMMERCE, CONSUMER PROTECTION, AND COMPETITIVENESS

OF THE

COMMITTEE ON ENERGY AND COMMERCE

HOUSE OF REPRESENTATIVES

ONE HUNDRED THIRD CONGRESS

FIRST SESSION

OCTOBER 5, 1993—TESTIMONY OF HHS SECRETARY SHALALA
OCTOBER 12, 1993—LABOR AND BUSINESS VIEWS
OCTOBER 14, 1993—CONSUMER, INSURER, AND PROVIDER ADVOCATES
OCTOBER 28, 1993—EXECUTIVE AGENCY TESTIMONY: IMPACT ON THE ECONOMY

Serial No. 103–75

Printed for the use of the Committee on Energy and Commerce

U.S. GOVERNMENT PRINTING OFFICE

77-929 WASHINGTON : 1994

For sale by the U.S. Government Printing Office
Superintendent of Documents, Congressional Sales Office, Washington, DC 20402
ISBN 0-16-044158-7

COMMITTEE ON ENERGY AND COMMERCE

JOHN D. DINGELL, Michigan, *Chairman*

HENRY A. WAXMAN, California
PHILIP R. SHARP, Indiana
EDWARD J. MARKEY, Massachusetts
AL SWIFT, Washington
CARDISS COLLINS, Illinois
MIKE SYNAR, Oklahoma
W.J. "BILLY" TAUZIN, Louisiana
RON WYDEN, Oregon
RALPH M. HALL, Texas
BILL RICHARDSON, New Mexico
JIM SLATTERY, Kansas
JOHN BRYANT, Texas
RICK BOUCHER, Virginia
JIM COOPER, Tennessee
J. ROY ROWLAND, Georgia
THOMAS J. MANTON, New York
EDOLPHUS TOWNS, New York
GERRY E. STUDDS, Massachusetts
RICHARD H. LEHMAN, California
FRANK PALLONE, Jr., New Jersey
CRAIG A. WASHINGTON, Texas
LYNN SCHENK, California
SHERROD BROWN, Ohio
MIKE KREIDLER, Washington
MARJORIE MARGOLIES-MEZVINSKY, Pennsylvania
BLANCHE M. LAMBERT, Arkansas

CARLOS J. MOORHEAD, California
THOMAS J. BLILEY, Jr., Virginia
JACK FIELDS, Texas
MICHAEL G. OXLEY, Ohio
MICHAEL BILIRAKIS, Florida
DAN SCHAEFER, Colorado
JOE BARTON, Texas
ALEX McMILLAN, North Carolina
J. DENNIS HASTERT, Illinois
FRED UPTON, Michigan
CLIFF STEARNS, Florida
BILL PAXON, New York
PAUL E. GILLMOR, Ohio
SCOTT KLUG, Wisconsin
GARY A. FRANKS, Connecticut
JAMES C. GREENWOOD, Pennsylvania
MICHAEL D. CRAPO, Idaho

ALAN J. ROTH, *Staff Director and Chief Counsel*
DENNIS B. FITZGIBBONS, *Deputy Staff Director*
MARGARET A. DURBIN, *Minority Chief Counsel and Staff Director*

SUBCOMMITTEE ON HEALTH AND THE ENVIRONMENT

HENRY A. WAXMAN, California, *Chairman*

MIKE SYNAR, Oklahoma
RON WYDEN, Oregon
RALPH M. HALL, Texas
BILL RICHARDSON, New Mexico
JOHN BRYANT, Texas
J. ROY ROWLAND, Georgia
EDOLPHUS TOWNS, New York
GERRY E. STUDDS, Massachusetts
JIM SLATTERY, Kansas
JIM COOPER, Tennessee
FRANK PALLONE, Jr., New Jersey
CRAIG A. WASHINGTON, Texas
SHERROD BROWN, Ohio
MIKE KREIDLER, Washington
JOHN D. DINGELL, Michigan (Ex Officio)

THOMAS J. BLILEY, Jr., Virginia
MICHAEL BILIRAKIS, Florida
ALEX McMILLAN, North Carolina
J. DENNIS HASTERT, Illinois
FRED UPTON, Michigan
BILL PAXON, New York
SCOTT KLUG, Wisconsin
GARY A. FRANKS, Connecticut
JAMES C. GREENWOOD, Pennsylvania
CARLOS J. MOORHEAD, California (Ex Officio)

KAREN NELSON, *Staff Director*
MICHAEL M. HASH, *Counsel*
ANDREAS G. SCHNEIDER, *Counsel*
HOWARD COHEN, *Minority Counsel*
MARY M. MCGRANE, *Minority Counsel*